T0406947

PORSCHE OUTLAWS

The shape may be familiar...but nothing here is the same. A traditonal hot rodder's twist on a '79 911 SC.

PORSCHE OUTLAWS

STUTTGART HOT RODS

MICHAEL ALAN ROSS

FOREWORD BY HURLEY HAYWOOD

Quarto.com

First Published in 2024 by Motorbooks, an imprint of The Quarto Group,
100 Cummings Center, Suite 265-D, Beverly, MA 01915, USA.
T (978) 282-9590 F (978) 283-2742

Motorbooks titles are also available at discount for retail, wholesale, promotional, and bulk purchase. For details, contact the Special Sales Manager by email at specialsales@quarto.com or by mail at The Quarto Group, Attn: Special Sales Manager, 100 Cummings Center, Suite 265-D, Beverly, MA 01915, USA.

28 27 26 25 24   1 2 3 4 5

ISBN: 978-0-7603-8263-9

Digital edition published in 2024
eISBN: 978-0-7603-8264-6

Library of Congress Cataloging-in-Publication Data

Names: Ross, Michael Alan, author.
Title: Porsche Outlaws : Stuttgart hot rods / Michael Alan Ross.
Description: Beverly, MA : Motorbooks, an imprint of the Quarto Group, 2024. | Includes index. | Summary: "Porsche Outlaws is the in-depth story of the Porsche enthusiast subculture known as "Outlaws" and the modified 356, 911, 912, 914, 924, and 944 cars commonly hot rodded"-- Provided by publisher.
Identifiers: LCCN 2024015344 | ISBN 9780760382639 | ISBN 9780760382646 (ebook)
Subjects: LCSH: Porsche automobiles--Customizing--History. | Porsche automobiles--Motors--Modification--History. | Hot rods--History.
Classification: LCC TL215.P75 R67 2024 | DDC 629.228/609--dc23/
eng/20240425
LC record available at https://lccn.loc.gov/2024015344

Design: Cindy Samargia Laun
Cover and Jacket Images: Michael Alan Ross

Cover Caption: In true hot rodder style, you really need to look closely at this 930 Turbo to understand the masterful work created by Rob Ida.
Front Endpaper: The attention to detail in John Oates's Emory Outlaw is exemplary of his passion. A wooden wheel for a guitar player makes perfect sense. MICHAEL ALAN ROSS
Back Endpaper: Perhaps the most feared Outlaw of all, Fred Veitch's one-off 356 "Gersetzlosser" (German for lawless). MICHAEL ALAN ROSS

Printed in China

CONTENTS

My father, John, took me to my
first car show in Palo Alto, California.
There, my love of cars exploded after seeing a
Kaiser Darrin, a Mercedes-Benz 300 SL Gullwing,
and a Cord convertible all in one place.
Wow! What a day.

John never drove a fancy car,
but he instilled in me at an early age his
appreciation for great craftsmanship and design.
He also gave me my first camera.

Thanks, Dad, it all worked out!
JOHN J. ROSS, 1920–2003

ACKNOWLEDGMENTS

A project like this does not come together without the support of amazing people.

First and foremost, my wife, Danielle, for putting up with countless hours away from home gathering images and putting up with endless questions like "how does this sound?"

The long list of friends who encouraged me to take on this project: Tom Cotter, this never would have happened without your inspiration and encouragement; Rob Gibby for showing me the power of networking and teaching me to simply ask for the business; Greg and Larry Ross for putting up with your nutty, little car-crazy brother; Glenn Chiarello for your friendship and consistent encouragement; Bruce Meyer for opening your doors and sharing your world and kindness; Carl Magnusson for taking a chance on a guy with a camera and a dream; Lyle Allan for those early years in the dark room and years of enlightenment ever since; Pete Stout for welcoming me to the fold; Randy Leffingwell for treating me as an equal from the start; Chris Harrell for having my back and believing in my talent; Behrouz Khanvali for always showing up when I needed him most; Mike Hinton for fact checking the hell out of things; David Mathews for being totally in sync with me every damn time; Richard Baron for ALWAYS making me look good; Rob Sass for letting me run with the ball; Vu Nguyen for never wavering in your commitment; Sean Cridland for simply being the calming force in the room; Ray Shaffer for your knowledge and willingness to share; Basem Wasef for treating me like a brother and knowing exactly what to say and when; Christian Garibaldi for guiding me through all the technical crap; Tom Neel for being of an artful mind; Zach Todd for your quick responses and "fixin' it in post;" Ed Taylor for your unyielding kindness; and Joey Shimoda, Sue Chang, and the entire team at Shimoda Design Group for their design support and creative genius.

Those who gave up their time and shared their wisdom and talents: Rod Emory and family, Patrick Long, Jeff Zwart, Carl Magnusson, Jim Goodlette, Jack Walter, Jay and Mark Wiener, Pete Ritter, Johnny & Gloria Riz, Bisi Ezerioha, Magnus Walker, Rob Ida, David Keens, John Oates, Gary Hustwit, Joshy Robots, Koichiro Kanda, Pete Stout, David Coleman, Jim Breazeale, Richard Breazeale, Mike Prstojevich, Jeff Higgins, Bob Aines, Freeman Thomas, Steve Hatch, David Miller, Joey Seely, Scott Birdsall, Kayla Delehant, Clay Carr, and Cameron Wayland.

If I happen to have left you off this list, please know I will be extremely embarrassed yet forever grateful for your support and kindness.

FOREWORD

A Few Words on Outlaws
By Hurley Haywood

Hurley Haywood is about to climb into the Porsche 962 C Le Mans car at the 1997 Monterey Historic Automobile Races. Porsche was the featured marque that year since it was their 50th Anniversary.

As a Porsche ambassador for over 50 years, I've raced dozens of the most successful racing cars ever to hit the track. I've driven several hundred Porsche demo cars as daily drivers. And, of course, I have had the opportunity to view thousands of enthusiasts' cars at various concours and cars-and-coffee events. I was even born the same year as the Porsche brand—1948—so my entire life has run parallel with the great cars and people who have made Porsche what it is and secured its place in historic automotive lore as one of the most prestigious brands in the contemporary marketplace.

Over the years I've seen a lot of trends in Porsche ownership, including converting various race cars from one model to another, updating and back-dating body kits, hot-rodding, and the recent obsession with perfect restorations and unrestored originals. But one of the most interesting subcultures that I've come across is the "outlaw" movement of recent years.

It's thought that the original "outlaw" Porsche 356 movement came from hot-rodder Gary Emory, though his son Rod has made a great business of it. In fact, I recently purchased a Rod Emory–built Speedster for my own collection. It's become my favorite car for touring, and it gets a lot of attention wherever I take it. Outlaws, as it turns out, are fun!

Since Gary and Rod's initial popularization of the 356 outlaw movement, it's moved in several new directions. Both Magnus Walker and the R Gruppe club of Southern California have taken the outlaw movement into the world of the early 911s with some fun and beautiful results, individualizing their cars so that they are both personal expressions of their own creativity while also harkening back to the club-racing days of the 1960s, 1970s, and 1980s. And while I appreciate those who work to create spot-on, perfectly executed original restorations, I give a wink and a nod to those whose cars are expressions of their own unique love for the brand. It's a big world and there's room for both.

Another niche in the outlaw movement now includes several builders and enthusiasts using the 911 chassis as the basis for off-road cars. Can you imagine Porsche 911s that are purpose-built to drive through deserts and bogs and get dirty inside and out? It looks like a lot of fun! Now Porsche itself has joined the trend with its own Dakar version of the 911.

And I can't leave out the exotic cars-and-coffee specials that I see all around the country. Who would have imagined tuner versions of the 997 and 991 that develop close to 1000 horsepower [735 kilowatt (kW)] and look absolutely stunning inside and out with shimmering titanium and carbon fiber trim pieces turning each car into a modern work of art.

All of these creations reach back to the one thing we've all had in common since the first Porsche roadster rolled out of the Gmünd sawmill in 1948: our love of a brand that exudes excellence, performance, and individual style personified. The outlaws continue to push our perceptions of what Porsche is, and I can't help thinking that somewhere in the heavens, Ferry Porsche is smiling.

—Hurley Haywood, 2023

PREFACE

In the Porsche enthusiast community, there are those who believe that cars should be preserved in their original "from the factory" condition, and there are those who challenge convention, reshaping their machines into something entirely new and thrilling. Welcome to the exhilarating realm of Porsche Outlaws—the subculture that defies tradition and embraces the wild, untamed spirit of hot-rodding.

Porsche Outlaws is a testament to the passion, innovation, and unbridled creativity of those who have dared to reimagine and modify Porsches of all kinds—not just the 356 but the 912, 911, transaxle cars, and even off-road vehicles like the Cayenne or Macan. Air-cooled or water-cooled, it's all fair game. This book also asks the question "Was the Porsche brand developed out of true hot-rod roots?"

We find ourselves at a unique intersection of automotive history and a growing movement that captures the imagination of gearheads around the world. The creation of this book is not merely about chronicling the past. It's also a celebration of the present and a glimpse into the future of automotive artistry.

Why now? The answer is simple. In a world that is increasingly defined by automation, electrification, and homogeneity, there is a burning desire for individualism and authenticity. Porsche Outlaws, with their unconventional approach to high-performance engineering, customization, and sheer driving pleasure, have emerged as an antidote to the mainstream. This book is an exploration of the contemporary resurgence of the Porsche Outlaw movement, where the classic marries the contemporary, where innovation meets tradition, and where the spirit of rebellion keeps the automotive world on the edge of its seat.

So, who am I to author this tale? Well, my camera has provided me with years of experience in and around hot-rod culture as well as an extensive background with Porsche, making me the perfect candidate to undertake this project. As an automotive photographer and author and a passionate automotive enthusiast, I have embarked on a journey to unearth the untold stories and unsung heroes behind the Porsche Outlaw phenomenon. Through countless interviews, extensive research, and an unwavering commitment to preserving the essence of this unique automotive subculture, I've had the privilege of delving into the world of those who dare to defy expectations, rewrite rules, and breathe new life into the iconic Porsche marque.

Within these pages, you'll find the stories of dedicated craftsmen, engineers, designers, and visionaries who have poured their heart and souls into their automotive passion. It's a testament to the tireless pursuit of perfection and a tribute to the Porsche Outlaws who have dared to dream and more importantly, dared to act upon those dreams. But the question is, "Exactly who or what is the outlaw? Is it the car or is it the person?"

Prepare to embark on a thrilling journey through the world of Porsche Outlaws, where speed, style, and audacity converge. Whether you're a devoted "Porschephile" enthusiast or simply an admirer of the extraordinary, I invite you to join me in celebrating the rebels, the rule-breakers, and the dreamers who have made the Porsche Outlaw movement an indelible part of automotive history.

Let's dive into a world that echoes with defiance and the only road worth driving is the one less traveled. Welcome to the realm of Porsche Outlaws, where the sky's the limit and the horizon is just the beginning.

—Michael Alan Ross

Everyone has a Porsche-that-got-away story. It was a sad day when I sold this car. I think of it often.

Don't let the suits fool you. Instead, imagine if you will, Dr. Ferdinand Porsche and his son Ferry in white t-shirts, jeans, and high top PRO-Keds just like any other So Cal family of hot-rodders. CORPORATE ARCHIVES PORSCHE AG

THE ORIGINAL STUTTGART HOT-RODDERS

In this shot, It's hard to deny the strong resemblance not only between family members, but the cars as well. CORPORATE ARCHIVES PORSCHE AG

When Porsche marked its 75th anniversary in 2023, it shared that milestone with another beloved enthusiast brand: Petersen Publishing's *Hot Rod Magazine*, which launched in 1948 as well.

Is it a coincidence? Almost certainly. But it's also a marker of the times. No matter which side of the Atlantic you were on, World War II had had a direct impact on the automotive industry worldwide and "hot-rodding," as a concept, was becoming mainstream.

War veterans worldwide returned home with a new set of skills and perspective. Many young men had gained hands-on experience and training that had expanded their horizons. Suddenly, the kid down the street could shape anything you wanted out of sheet metal. His buddy down the block had acquired an electrical engineering degree, and that kid's older brother had served in the motor pool and had learned all kinds of tricks to make internal combustion powered things faster. The performance engineering techniques brought back from the field (literally) are still applied today.

The Porsche pedigree can be simplified as follows: In 1946, Dr. Ferdinand Porsche, who had been imprisoned for 22 months on war crimes charges, was released in exchange for money paid by Italian sports car manufacturer Cisitalia. To reciprocate and repay the debt, Ferdinand agrees to design and build three race cars for Cisitalia. From there, he and his son Ferry go on to design and build the sports car of his dreams in a barn in Austria using Volkswagen (VW) and other parts. This initial 356/1 was the catalyst for the creation of the new company we know today as "Porsche."

So, a fair question arises: Were Ferdinand and Ferry Porsche hot-rodders? It would surely seem so. Anyone who builds their own car out of spare and available parts is definitely a hot-rodder. With that said, does this make Porsche number #001 a hot rod? If that's true, can we assert that the foundation of the Porsche company was built upon a hot rod?

LEFT: Shop discussions with one of the original wooden bucks for the 356 Coupe. CORPORATE ARCHIVES PORSCHE AG

BELOW: This workshop in Gmünd could be on any countryside or back alley in Anaheim, CA, but it's what came out of it that built an empire. CORPORATE ARCHIVES PORSCHE AG

We know Porsche as the great sports car company that also manufactures everything from sport utility vehicles (SUVs) to hybrid and full electric sedans for today's market. But when you start to break down the history of the brand and the cars they have designed and raced over the decades, it's impossible to deny the hot-rodding DNA that exists in every Porsche.

Many of Porsche's most iconic designs have been derived from race cars, and aren't those race cars hot rods? When you modify a stock Porsche 356 to go racing at the 24 Hours of Le Mans in 1951, you must make it faster, more aerodynamic, and more robust. This sounds a lot like building a hot rod to go racing at the strip or on the dry lakes. What's the difference between racing on the street and racing on a road course sanctioned by a racing organization? The only difference is money. Technology and ingenuity make a hot rod, and what makes a car a *hot rod* is a *hot-rodder.*

TOP LEFT: The simplicity of Porsche #001 is the definition of the company's favorite build theme: "form follows function."

TOP RIGHT: Often overlooked is the detail left of the steering column. The ignition placement there has been part of Porsche design since day one. Many suggest this is the preferred placement based upon the famous "Le Mans Start."

OPPOSITE: The minimalist approach follows through into Porsche #001's interior. Also note the one of a kind slotted body panels offering ventilation for its mid-mounted engine.

I asked former Porsche Senior Designer Freeman Thomas if he thought the Porsche family were hot-rodders:

> **"Ferdinand Porsche had that quality which became an ultimate hot-rodder. He learned so much from aircraft as well as a great deal about mechanical fuel injection. He learned about engine technology, supercharging, and making things lightweight. Look at the displacement [and engine configuration] of the early Auto Union cars. Yeah, that apple didn't fall far from the tree, and when you think about the first Porsche under Ferry Porsche, it was a hot-rodded Volkswagen. It had twin carbs on it, and the first one was mid-engine, then rear-engine. They were changing things every year. They started building their own [engine] blocks—real hot-rodding.**
>
> **Come to think of it, when I was at Porsche I worked with Tony Lapine. Tony was an engineer [who] came to Porsche fresh out of GM's Skunk Works . . . another hot-rodder!"**

The conclusion is inevitable: Porsche is rooted in hot-rodding. Consider the trickle-down effect and how technology derived from racing makes production cars safer, faster, and more efficient no matter what you are driving. From a GT3 RS to a base Panamera or Macan, a production Porsche outhandles and outperforms just about any car you can imagine.

When you consider that the Porsche legacy began with one car in a barn in Austria and developed into one of the most successful car companies of all-time, well that's a pretty good result for a couple of hot-rodders. And it's a clear inspiration for Outlaw Porsches.

33115
140
160
20
40
10
50
60
U/min
x100

The Porsche 550 Spyder was renowned as a "giant killer" based on its racing success throughout Europe and Mexico against larger displacement competition.

2

BLURRED LINES

James Dean driving in his famous "Little Bastard." ALAMY STOCK

Over its history, Porsche has created cars that might be thought of as "blurred lines." These were cars that took a significant step away from their production-car basis or were created from scratch. These cars stepped out of the box. Maybe they even kicked the box over. Some of these historic outliers went on to become actual production cars, though unlike any other cars in Porsche's lineup. Frequently, these cars stemmed from race cars, concept cars, or even special design gifts created for the Porsche family. The most significant of these cars inspired the hot-rodders of the Porsche Outlaw movement and deserve examination.

550 SPYDER

We've all heard the expression that inside every Porsche is a race car. Our first example is exactly that. In 1953, Porsche developed the 550 Spyder, which made its debut at the Paris Motor Show that same year.

Porsche built the 550 Spyder from 1953 to 1956. During that time only 90 were constructed, and each car was customized for the individual owner based on where it would be raced and who would be piloting it.

One of the 550's most memorable results was Hans Hermann's victory in #55 in the under-1500cc class of the 1954 La Carrera Panamericana.

The Spyder wore an all-aluminum body mounted to a flat-welded steel frame. It was powered by the legendary Fuhrmann engine (Type 547), which was mounted midship directly behind the interior firewall and ahead of the rear axle. Ferdinand Porsche had pioneered this mid-engine design and layout with the Auto Union Grand Prix cars of the 1930s. The design works well because it centers the weight of the vehicle and provides neutral handling. Derivations of 550's basic design would be seen in multiple Porsche racing and production cars including the 904, the 917, the GT1, Carrera GT, 918 Spyder, and the latest Porsche 963 race cars. Tragically, the 550 may be best remembered as the car in which actor and racer James Dean met his demise on California State Route 46 (then U.S. Route 466) when he collided with a Ford Tudor.

ABOVE: Though not the car pictured, of course, perhaps the most famous 550 Spyder was the one owned by actor James Dean. Unfortunately, it was the car he died in on September 30, 1955, in Cholame, CA, after a crash with a 1950 Ford Tudor.

LEFT: The 550 Spyder was followed by the RSK, which can be easily recognized by its unique larger intakes on either side just behind the doors.

ABOVE: The Porsche 904 GTS, a.k.a. Porsche Carrera GTS, was an artful masterpiece and Porsche's first venture into the use of fiberglass for body construction.

OPPOSITE, TOP: The Porsche 904 GTS profile is unmistakable and unlike any Porsche before or since.

OPPOSITE, BOTTOM LEFT: Note the use of aircraft-style doors that mold into the 904 roofline.

OPPOSITE, BOTTOM RIGHT: The elongated plexiglass headlamp covers were used on both race cars and street cars.

904 CARRERA GTS

In anticipation of the 1964 racing season, Porsche introduced the 904 in late 1963.

For the first time, Porsche did not construct an aluminum body for its race car, instead using a fiberglass-reinforced plastic body developed in collaboration with the Heinkel Flugzeugbau aircraft manufacturing company.

There were a total of 106 four-cylinder powered production cars created, retailing for $7,245 USD (US dollars). A six-cylinder version was also offered, with only a total of 40 constructed.

The factory also built three race cars fitted with flat eight-cylinder power plants. Those 225 horsepower (165 kilowatts [kW]) engines had been developed originally for the 1962 model 804 F1 car. Unfortunately, they were known for exploding flywheels.

The 904 is significant for one special reason: Its designer was none other than Ferry Porsche's son Ferdinand Alexander "Butzi" Porsche, renowned for his involvement with the creation of the 911 as well as his launch of Porsche Design. Butzi frequently described the 904 as his favorite work.

PORSCHE

This side view gives you a clean view of the stance, vented flexi windows (both vent widow and rear quarter), and the big picture of how much cleaner the car is with its tucked in bumpers. Look closely at how the rear tires tuck under the right rear quarters.

1967 911 R

Porsche's experience in rally and endurance racing is the foundation of its success, and the 911 R is a perfect example. In 1967, Ferdinand Piëch, Ferry Porsche's 29-year-old nephew, helped Porsche remain competitive at the highest level by creating the 911 R. The legendary R became part of every Porsche's DNA. It's the car that spoke to purists then, and it will continue to do so well into the future. You can't look at a new GT3 and not think of the 911 R.

The concept for the 911 R was to develop a car to push the boundaries of motor racing by deploying all that Porsche had learned to that point. That *R* stands for *Rennsport* (*racing* in German). Only twenty-three 1967 911 Rs were built, including the four prototypes, its rarity ensuring it would become an icon of icons. Aerodynamic enhancements, lightweight construction, and sharpened performance set the template still evident in today's Porsches.

The key features of the 1967 911 R that distinguish it from the 911 S are as follows:

- **Performance**: The 911 R's power source is none other than the air-cooled flat-six developed for the 906 Carrera 6 race car, producing 210 horsepower (154 kW) backed by a five-speed transmission.
- **Aerodynamics**: The differences between a 911 R and a stock 911 S are so subtle that it takes a trained eye to detect. Walking around the car, you'll notice things like the removal of the right outside mirror, molded bumpers, slightly flared fenders to accommodate wider tires, horn grills removed to increase airflow to accommodate larger oil coolers, through-the-hood gas filler placement, ventilated side windows, and smaller molded taillamps and front marker lamps.
- **Weight reduction**: Weight reduction was a key factor in the 911 R's success. Fiberglass panels were used for the doors, bumpers, engine cover, and hood. The interior was gutted, leaving behind only essential items, and what remained was either altered in some way or replaced with a lighter construction material. All insulation and sound deadening materials were removed. Metal door handles were replaced with leather straps. Metal hinges were drilled out or replaced with plastic parts. Even the instrument cluster was reduced to three gauges rather than five. The windshield glass was removed and replaced with 4 millimeter (0.2 inches) plexiglass, side glass was removed and replaced with Perspex, and the rear window was also removed and replaced with plexiglass measuring just 2 millimeters (0.1 inch) thick.

The 1967 911 R remains the lightest 911 from that period ever built, with a power-to-weight ratio matched by no other 911.

Undergoing its first tests in Weissach, the 911 R proved just 12 seconds slower than the Group 4 906 Carrera 6, assuring the car would be one of the most competitive GT cars of its time.

The performance-oriented 911 R became the benchmark for Porsche's dedication and commitment to motorsports. When Porsche created *The Sports Purpose Manual* years later, it was all based on the 911 R, which is also the holy grail for the original Porsche Outlaws and the R Gruppe. All hail the once and forever king.

BELOW LEFT: The bare bones interior of the 911R accentuates the theme of less is more and lighter is faster. The lack of trim, glove box door, and not even a radio blank cover really drive it home.

BELOW RIGHT: Perhaps the most replicated of features from the 911R to be used on Outlaw builds: These smaller and lighter dual tail lamps rather than the production cars wraparound chrome trimmed ones.

Both the side body script Carrera and Duck Tail rear spoiler were option delete items for the 73 911 RS. Trying to find one without both will be an endless search as they became the defining items of this icon.

1973 CARRERA RS

For Porsche enthusiasts old enough to remember Marx Brothers skit "Why a Duck?" from the 1929 film *The Cocoanuts*, the 1973 Carrera RS has particular significance.

It all goes back to the German Patent Office and document No. 2238704 filed by Porsche AG requesting a patent for a passenger car with a rear spoiler or aerodynamic device designed to increase the downforce on the rear wheels to improve handling. Said device is what enthusiasts refer to as the *ducktail*.

Porsche's ducktail, designed by Hermann Burts and colleagues Tilman Brodbeck and stylist Rolf Wiener, made its debut on the 1973 Carrera RS and was one of many firsts unveiled on that car at the Paris Motor Show in October 1972. For one, the word *Carrera* had never been used in conjunction with 911 until then. Porsche chose the name as an homage to La Carrera Panamericana, a road race held in Mexico for five consecutive years in the 1950s (1950 to 1954). In 1953, Porsche earned a class victory with the 550 Spyder and an overall third place the following year.

Porsche designer Harm Lagaaij loved the idea of using the Carrera name so much that he had the graphics team create the iconic side stripes that run between the wheel arches and spell out the name in large letters. Tying the theme together, Porsche matched the spoke color of the Fuchs wheels to the side stripes and continued around the car with an accent strip in the same color on the front and rear bumpers. Versions of this striping have been used on several cars since and always signifies performance.

Just like its predecessor, the 911 R, the Carrera RS utilized lightweight construction evident in its molded bumpers. The difference was that the Carrera RS also had a front air dam as part of the bumper assembly, making it the first production car to ever have a front and rear spoiler.

Unlike the 911 R, the RS used stock taillamps, front turn signals, and marker lamps. The interior was less hard-core and more comfortable with sound deadening, creature features like conventional door handles, and a leather, padded dash making the car much more manageable as a daily driver.

But what may have seemed, comparatively, as a softening of the car's performance intent was offset in other ways. For example, the RS was the first time Porsche built a series production car with staggered wheel sizes, front to back.

This decision was directly handed down from Porsche racing engineers. Running wider wheels and tires out back required the development of the RS's famous wheel flares to accommodate the wider rubber within the body lines for aerodynamic purposes. These flares widened the body 42 millimeters (2 inches) at the rear. The resulting stance is legendary and a clear inspiration to outlaw builders.

The more aggressive stance was abetted by a suspension tuned for a stiffer ride via lighter and thicker anti-roll bars, reinforced control arms, and reinforced cross-members in the rear end.

The end result was a badass Porsche street racer powered by a 2.7-liter (164.5 cubic inch [cu in]) air-cooled flat-6 with 210 horsepower (154 kW) on tap that you could still use as your daily driver. It was the perfect, aggressive street car during the work week and an amazing track car for the weekends. Now, that's an *outlaw.*

The 73 911 RS came in two versions: Lightweight and Touring. Only 200 Lightweights were built (plus 17 homologation cars) in comparison to the 1,308 Touring versions. With a whopping weight difference of 220 pounds (100 kg), you can start to see the subtle differences here. Take a look at the bumpers guards, bumpers, and hood closures.

From this angle, you can see the cross sectioning of a 993 and a dune buggy of the Panamericana. CORPORATE ARCHIVES PORSCHE AG

PORSCHE PANAMERICANA

1989 was to be a momentous year at Porsche as it would mark Ferry Porsche's 80th birthday. To commemorate his eighth decade, Porsche's design team dreamed up something different as a birthday gift. Together, Harm Lagaaij, Steve Murkett, and Ulrich Bez worked together to create the Porsche Panamericana concept car. Its name was derived from the famous La Carrera Panamericana Mexican road race. The design team wanted to create a vehicle that could do just about anything the driver asked of it. Starting with a 1989 911 (964) Carrera 4 Cabriolet donor car, they created a radically redesigned body built of plastic and carbon fiber panels.

The unique cutaway wheel arches were designed to make suspension and wheel and tire changes easy in case Ferry decided to go off the beaten track.

The car also featured a unique roof, which could be configured many different ways, making the car a combination of a convertible, Targa, and Coupe. Studying the car, you begin to see styling cues that eventually surfaced in the Porsche 993. As an added feature, the Porsche crest was etched into the tires' tread pattern and mounted on unique (Panamericana only) Speedline three-piece rims.

Despite the design team's hopes for a possible limited run of this unique Porsche, only two were created. One was gift wrapped and presented to Ferry Porsche on his 80th birthday, while the other was unveiled at the 1989 Frankfurt Motor Show and then was shown at the Tokyo Motor Show.

TOP: The open air roof of the Panamaericana is totally unique. It's not a Targa or a Cabriolet.
CORPORATE ARCHIVES PORSCHE AG

BOTTOM: The Porsche family at the presentation of the Panamericana.
CORPORATE ARCHIVES PORSCHE AG

The Porsche 959 was something of a test mule for what would emerge from Zuffenhausen in the following years. It was basically the precursor to the 996 Turbo. The 959's unique raked headlamps are reminiscent of those later found on the Porsche 911 GT1 ('96).

PORSCHE 959

The Porsche 959 was produced from 1986 to 1993 as the road-legal production design vehicle built to satisfy Fédération Internationale de l'Automobile (FIA) homologation requirements for Group B rally cars. This meant Porsche had to build 200 street-legal cars. It also explains why the 1983 concept car was called the Porsche Gruppe B Concept and not the 959.

When released, the twin-turbocharged 959 was the world's fastest street-legal production car with a top speed of 197 miles per hour (mph) (317 kilometers per hour [kph]), with certain variants maxing out at 211 miles per hour (349 kph). The car was rugged and lightweight, constructed of aluminum and aramid (Kevlar) composite for body and chassis construction. For further weight savings, the floor pan was made from Nomex.

The 959 body included flared fenders front and rear, boxed rockers, and a one-piece rear wing molded into the deck lid.

The 959's all-wheel drive system was developed by Porsche-Steuer Kupplung (PSK) and gave the 959 unsurpassed adaptability and a competitive edge both as a race car and a *super* street car.

The road version of the 959 debuted in 1985 at the Frankfurt Motor Show and was available in both sport and comfort trims and priced at $225,000 USD. The manufacturer's suggested retail price (MSRP) was exactly half of Porsche's cost per vehicle. Production ended in 1988 with a total of 292 road cars built, though in total 337 cars were constructed, including 37 prototypes and preproduction models.

Between 1990 and 1993, Porsche built eight additional cars out of spare parts from their inventory in Stuttgart. All eight were comfort versions—four painted Guards Red and four in Arctic Silver. Based on limited supply and high demand and the fact these cars also featured a newly developed speed-sensitive damper system, the 959 became the most desirable model Porsche had offered to that point.

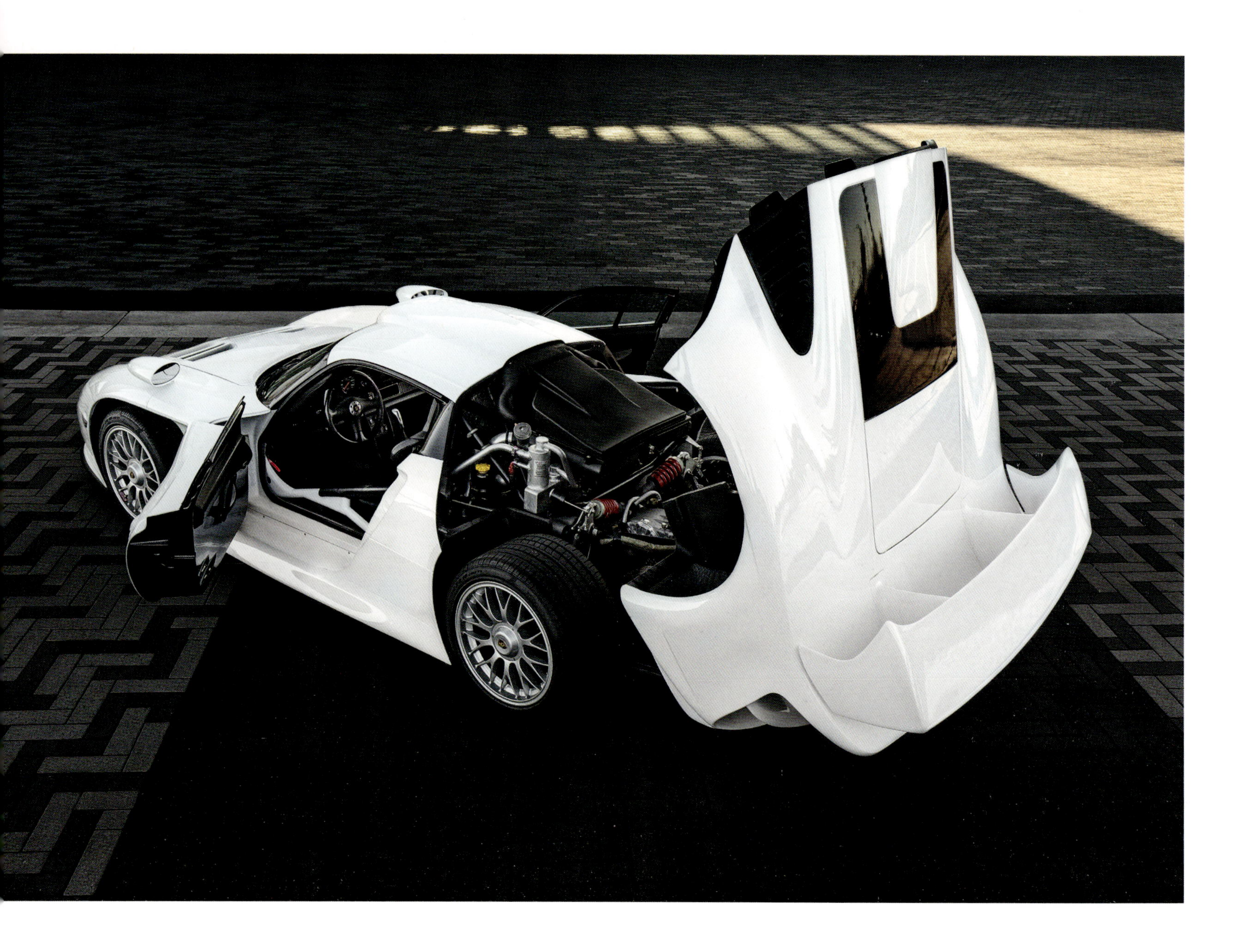

Easy access to this mid-engine Le Mans winning masterpiece is created by the one piece tilt-back rear clam shell body panel. Kevlar and carbon fiber construction made the massive combo of body and aero easy to maneuver.

911 GT1 STRAßENVERSION

This Porsche hot rod is the only car in the Porsche lineup ever to be named after the series in which it competed: GT1. FIA World Endurance Championship (WEC) GT1 class requirements dictate that a total of 25 street cars must be constructed to satisfy homologation regulations. The cars were designed in-house by Tony Hatter who had also designed the 993. In early 1996, the first car was delivered to the Federal Ministry of Transportation for compliance testing. The second vehicle was delivered to a collector in Bahrain. These first two GT1s are unique as they were equipped with 993 front headlamps.

The pure lines of the GT1 become even more fluid here in motion. You can almost see how it uses the air it slips through for cooling and downforce as it goes by.

Production cars are known as the 911 GT1 Straßenversion (Strassenversion). Approximately 20 units were completed in 1997. The newer version of the vehicle is easily spotted by its 996 front headlamps.

All but three of the GT1 Straßensversion were finished in either Arctic Silver or Fern White. The remaining three were Polar Silver, Indian Red, and Pastel Yellow. The first two versions of the car were competitive at Le Mans, but both fell short of Porsche's goal of winning outright.

Falling short of a goal is not something Porsche is accustomed to. So, in 1998, it produced one GT1-98 Straßensversion for homologation purposes and then created the 911 GT1-98 race car. This new race car proved to be exactly what Porsche needed to win Le Mans overall in 1998.

ABOVE: This extreme example of what's capable from Porsche Classic was unveiled at the PCA Werks Reunion at Amelia Island in 2019. The list of upgrades and custom features, including PTS Oak Green Metallic, would double the pages of this book.

OPPOSITE: Although the color GT Silver Metallic first appeared on the Carrera GT, some still went with Paint to Sample (PTS) colors. This particular car was done in Polar Silver. The standard colors included Guards Red, Fayence Yellow, Black, Basalt Black Metallic, GT Silver Metallic, and Seal Grey Metallic.

CARRERA GT

At this point, you're probably seeing a pattern of Porsche design trickle-down. The Carrera GT is another combination of previous designs involved with homologation and rules set by the FIA and Automobile Club de l'Ouest (ACO).

We can trace the roots of the Carrera GT back to the 911 GT1 and the LMP1-98. In 1998, there were some rules changed for any cars intended to race at Le Mans in 1999. Porsche's design was scrapped and never raced. That car, instead, became the Carrera GT. Initially designed for a turbocharged flat-6, the car was reconfigured to accommodate a V-10 originally designed and built secretly for a Formula 1 (F1) team before being shelved in 1992. It was resurrected again for the Le Mans prototype, the 1999 Porsche 9R3, and finally appeared in the Carrera GT in 2004. Essentially, one of the greatest Porsches of all-time is a mash-up of previously abandoned projects designed initially for racing.

Though the Carrera GT was first seen as a prototype at the 2000 Paris Motor Show, it would take revenue generated by Cayenne SUV sales to help fund the car's production.

The Carrera GT was the first Porsche production vehicle to use a carbon-fiber monocoque subframe. Carbon fiber panels were also used for select body panels.

Priced at $440,000 USD, the V-10 supercar was rated at 612 horsepower (450 kW) in its US guise and was initially available in six colors: Guards Red, Fayence Yellow, Black, Basalt Black Metallic, GT Silver Metallic, and Seal Grey Metallic.

This was the first street car Porsche ever produced with center-locking wheels. Because the wheel hubs were unique side to side, they are differentiated by color. On the driver side, the locking nuts are red, and on the passenger side, the locking nuts are blue.

ABOVE: Porsche knows how to design a supercar that lasts the test of time. Unlike previous cars, the 918 was Porsche's first consumer hybrid since the Lohner-Porsche Electomobile, which was debuted at the 1900 Paris Exposition.

OPPOSITE: From above we see how similar the 918 is to a 911 from the A-Pillar forward. We can also see the massive changes from the B-Pillar back.

918 SPYDER

Porsche's first production hybrid supercar was the 918 Spyder, which began production on September 18, 2013. The 918 Spyder was sold out by December 2014.

Shown as a concept at the Geneva Auto Show in March 2010, Porsche received 2000 declarations of interest. Porsche also unveiled the RSR version of the 918 Spyder, utilizing the 997 GT3 R hybrid technology, at the same show. However, the 918 RSR did not make it to production.

The 918 powerplant consisted of a V-8 engine developing 599 horsepower (441 kW) assisted by two electric motors delivering an additional 282 horsepower (207 kW) for a combined output of 875 horsepower (644 kW). All of this power was directed through a seven-speed gearbox utilizing Porsche's PDK dual-clutch transmission.

California, falling in line with current trends of course, was ahead of the curve and embraced the 918 Spyder with a total of 297 units sold.

For an additional $84,000 USD, you could check the Weissach Package box. This included BBS magnesium wheels, an extended rear diffuser, and an Alcantara interior. The package also contained various carbon-fiber parts including the windscreen frame, roof, rear wing, and a rearview mirror.

Porsche took a Weissach package–equipped 918 Spyder to the famed Nürburgring for testing. The results were phenomenal. Porsche factory driver Marc Lieb posted a then record-setting lap time of 6 minutes, 57 seconds at an average speed of 179.5 kilometers per hour (111.5 mph). It beat the previous record by a whopping 14 seconds and was the first road-legal production car to shatter the 7-minute barrier of the Nordschleife (the North Loop).

BOTH: Mission X is Porsche's first supercar designed with Environmental Protection Agency (EPA) regulations in mind. The current trend towards electric vehicles (EVs) has created some challenges regarding successful market penetration. Creating excitement for a car that doesn't make noise is still an obstacle . . . creating a supercar is even harder.
CORPORATE ARCHIVES PORSCHE AG

MISSION X

On June 8, 2023, at the Porsche Museum in Stuttgart-Zuffenhausen, to celebrate 75 years of Porsche sports cars, Porsche debuted their newest concept car—the hypercar for the future called the Porsche Mission X. The car is similar in dimension to the Carrera GT and 918 Spyder, with a wheelbase of 107.4 inches (2.73 m). And like the Porsche 917, it is equipped with Le Mans–style doors and Daytona windows. According to The Porsche Newsroom US:

> **"The Porsche Mission in the X is a technology beacon for the sports car of the future. It picks up the torch of iconic sports cars of decades past like a 959, the Carrera GT and the 918 Spyder before it, the Mission X provides critical impetus for the evolutionary development of future vehicle concepts," says Oliver Blum, Chairman of the Executive Board of Porsche AG. "During the dream and dream cars are two sides of the same coin for us: Porsche has only remained Porsche by constantly changing."**
>
> **Michael Mauer, Head of Style Porsche states, "The concept study symbolizes a symbiosis of unmistakable motorsport DNA with a luxurious overall impression."**

PORSCHE

The unmistakable nose of the Dean Jefferies' *Kustom Karrera* is as unusual now as it was back in 1957. The car has never been replicated.

3

THE ORIGINAL OUTLAW

Hand painted by Dean Jefferies and clearly stated in quotes on the rear of James Dean's Porsche 550 Spyder are the words *Little Bastard*. Apparently, this pronouncement was self-referential on Dean's part. ALAMY STOCK

DEAN JEFFRIES

You can't talk about outlaws and not talk about the connection between Dean Jeffries and James Dean. They were both into Porsches and regularly attended race events together. In 1955, James traded in his Porsche Speedster for a brand new 550 Spyder. He drove the car to Jeffries' Lynwood, California, shop and asked him to paint the number *130* in flat-black, washable paint on the hood, doors, and rear deck lid. James was prepping the car to go racing that weekend in Sacramento. The plan was for Jeffries to join him, but he had too much work to do in the shop and opted out.

While James was at the shop, Jeffries remembered him once referring to himself as a "little bastard," and he suggested James paint that moniker on the back of the car. James loved the idea, and Jeffries later added that epithet at George Barris's shop, Barris Kustoms. The rest, as they say, is history.

Two years later, in 1957, a young man pulled up to Barris's shop where Jeffries was working and asked if Jeffries wanted to trade cars with him. Jefferies had a 356 1500 Super at the time and the car offered in trade was a Carrera Coupe. Jefferies agreed, and they exchanged pink slips on the spot.

Apparently the young man's father had bought the car for him while on a trip. The kid really didn't know what he had. He had traded a four-cam Carrera straight up for a Porsche 356 1500 Super! It was an excellent outcome for Jeffries.

A week later the kid's father showed up at Barris' shop furious and wanting to get the car back. Jeffries wouldn't budge. It was his car now.

At the time, there was a Mercedes-Benz 300 SL in the Barris shop to which Von Dutch (Kenneth Robert Howard) was applying a flame paint job. Jeffries fell in love with certain visual cues on the 300 SL and worked to apply them to his Carrera. Thus, it came to pass that in the back of Barris's shop in 1957, Jefferies went to work creating what would become the first Porsche 356 Outlaw. His car incorporated several European touches like frenched fog lamps and headlamps as well as a significant redesign to the front end.

Jeffries also redesigned the rear deck lid vent area, taking cues from the Gullwing 300 SL and creating his own version of that car's roof rear vents above the back window. He then redesigned the rear taillamps to resemble those of the Mercedes as well.

Jeffries was perhaps best known as a master painter, and his specialty was pearlescent paints. He painted his Carrera pearlescent silver. Jeffries also applied a special trick he had to make paint last. Most painters applied a clear coat finish to their cars, and within a few years, that coat would start to yellow and chip. Jefferies used an aviation clear coat finish that had a higher UV factor, and the coat he applied never changed.

One of the earliest images of the freshly completed 1956 Carrera Outlaw with Dean Jeffries. JACK WALTERS COLLECTION

Here's Dean Jefferies with his work of custom-car art, the *Mantaray*, which was based on a pre–WWII Maserati and was inspired by and named after the "devil of the sea," the manta ray.
JACK WALTERS COLLECTION

When the Carrera was complete, Jefferies took it to numerous events and drove the wheels off it. Even in the late 1950s, Porsche people weren't happy about the modifications, but now people embrace the car—including members of the Porsche family.

Jeffries enjoyed the car for several years, until the day a man walked into his shop and offered to buy it, cash in hand. They came to an agreement, and the car was gone. A few weeks later, Jeffries was confronted by two FBI agents who wanted to know all about the Carrera and how much money he had been paid for it. Jefferies told them it was none of their business and sent them on their way (an outlaw move in itself!). He was later distressed to learn that the man who had purchased his car was a bank robber that had been on the FBIs Most Wanted list for his latest heist in December 1961 that ended with the death of a security guard—that's where the cash had come from.

The FBI finally caught up with this entirely different sort of outlaw in Florida where the car was seized and eventually sold at auction. Whoever purchased the Carrera at auction eventually traded it in at a dealership near Orlando, Florida, where it was sold from the used car lot. Thus began part two of the Jeffries Outlaw's long, strange trip.

JACK WALTER

In 1970, Jack Walter was 19 years old, living in Atlanta, Georgia, and in need of a car. He had found a cool Porsche 356 Speedster, and he had saved a total of $400. Unfortunately, Walter's friend, Jim Downing, wanted $600 for the car. In an effort to get enough money via a parental bridge loan, he asked his father to check the car with him to see if they could work out a deal for the extra $200. One look at the car and Walter's father said, "There's no way you're gonna buy that car!"

So Walter ended up with a Chevrolet Corvair convertible. The motor was shot, but it was turbocharged. He rebuilt the engine, tweaking it to 250 horsepower (184 kW). It was a hot rig by any measure, but Jack still wanted that Porsche—or any Porsche, really.

Walter's best friend, Paul Doale, had moved to the other side of Atlanta. Every now and then, Walter would stop over to hang out. One day, he pulled into the driveway and there sat this funny-looking Porsche. He got out of his Corvair, circled that Porsche a few times, and then knocked on his friend's front door. As it turned out, the Porsche belonged to his friend's older sister, Margaret. She knew how he felt about Porsches, so she asked Jack teasingly, "What do you think of my new car?"

"Sell it to me," was Walter's immediate reply.

About a year later, Margaret had decided that she wanted to go to Nepal. She was raising money for her trip, and she called Walter. "You still wanna buy my Porsche?"

BELOW LEFT: Jack did a great job fixing the car's nose and added his own detail with his pointer racing stripe. JACK WALTERS COLLECTION

BELOW RIGHT: One of Jack's prized possessions: A photo of his car with the pointed stripe signed by George Barris in 1980 along with George's business card and a signed business card form Dean Jeffries dated 1995.

"Haven't I been trying to buy it from you for a year?" he asked. "Just tell me how much you want." The sum of $1,100 bought Walter's dream car that very day. His dad remained skeptical.

The Porsche's nose was punched in a bit, but Walter didn't care. He had it fixed within the year, albeit filled with Bondo, and painted white. Deciding it looked a little plain, Walter applied a blue stripe up the middle of the car to give it a racier look.

In the early 1970s, Walter started taking his car to Porsche events in the Atlanta area. Other owners seemed horrified. There were no modified 356s. Nobody was doing *that*, and the term *Outlaw* did not exist, at least in a Porsche context. But according to Walter, "Gary Emory says he saw the car at a show in 1971 and it was the car that inspired him to start modifying Porsche 356s."

Walter had the car in a little shop called Auto Mod in Sandy Springs, Georgia, where he was having a mirror fixed and his driving lights adjusted. While the mechanics were fiddling with his car, a man passing the shop approached him and said, "Hey, I know that car. I saw that car in George Barris' shop in 1957." This was Walter's first clue regarding the origins of his car. The passerby didn't recall any specifics, as he was just a kid when he'd seen it.

A few months later, Ray Ringler, who was the head of the local Porsche Club of America (PCA) region in the early 1970s, brought Walter a copy of the October 1959 issue of *Rod & Custom Magazine* featuring his car on the cover. The article

Here's the original and freshly rebuilt four-cam motor as it arrived from Europe. JACK WALTERS COLLECTION

explained that the cover car had been built and owned by Dean Jefferies while he was working for George Barris. The title of the article was "Silver Satin," which Jack thought was the name of the car. As it turned out, the car was named *Kustom Karrera*. Finally, he had solved the mystery of who had built his car and why it was so unusual.

Now, the whole thing made sense. Dean Jeffries was not only a painter. He was also a metal fabricator, and he was building movie cars. Walter worked for Lockheed from 1980–2015, so every time the company sent him from his Atlanta base to California, he made a point to stop by Dean's shop. The two became fast friends, bonding over the *Kustom Karrera*, which they jokingly referred to as the *Killer Karrera*. They spent hours together, Jeffries answering Walter's questions about the car as the latter worked to restore it.

Consider the car's dashboard, which Walter assumed was engine turned. Nope. Instead, Dean had applied silver leaf to the metal and then taken a cotton ball dipped in black paint and swirled it over the silver leaf. It was then topped with a clear coat. "Like an idiot, when I got the car, I couldn't figure it out," recalled Walter. "So I sanded [the dashboard] down, and I painted it black. How did I know?" Jeffries taught him how to recreate the original finish.

As restoration progressed, a key element loomed in Jack's mind. The car's original four-cam motor had been missing from the day he had bought it. Who knew where *that* ended up? But Walter understood that he needed the correct engine. Jack called his friend Ernie Cabrera, who said, "I know somebody in Florida who's got some." Bingo!

ABOVE LEFT: Reunited at The Amelia Island Concours 2011. Dean Jeffries, Jack Walter, and the Karma Karrera.

ABOVE RIGHT: The four-cam sits in the car now with all the "shiny bits" in place. This is probably the first and last time you will see this much chrome anywhere near a four-cam.

NEXT PAGE: Dean Jefferies would be proud to see his car restored to the level accomplished by Jack Walters. Here it is on the Petersen Automotive Museum's roof in Los Angeles moments before it was placed on display in 2023.

Carrera

So in 1975, after a few phone calls, Cabrera was on his way to Jacksonville, Florida, from Atlanta, Georgia, in his Honda Civic hatchback to retrieve a four-cam for the *Kustom Karrera.*

Jack loves to tell the story about Cabrera's drive home up I-95 with that four-cam in the back of his Civic. "Somewhere just outside of Jacksonville, Ernie was passed by a Mercedes SL convertible, with the license plate HHHHHH, moving at quite a clip. Ernie then noticed the Mercedes get on the binders and slow down until they were door-to-door. Much to Ernie's surprise, he looked over to the Mercedes [to see] Hurley Haywood giving him a thumbs up. It only lasted a moment before the Mercedes was back on the gas."

Once Walter had the four-cam in his workshop, he bore-scoped it and found it had a hole burned in one of the pistons. The motor would need to come apart.

This four-cam engine had a very low serial number, and it also had a side port for oil connection tubes that were used only on the 550 Spyder. Yes, the four-cam Ernie had unearthed was a 550 engine and not a 356. In his definitive book on Carrera engines, Steve Heinrichs includes a chart documenting all the four-cam engine numbers and which car they were installed in. The number on Jack's engine was 90009, making it the ninth four-cam engine ever built. It was originally in 550 Spyder number 0013.

Walter posted on the Porsche 356 Registry website asking if anyone knew where 550 0013 was. Someone responded stating that they knew the car's owner, and Walter was put in contact with him. Shortly thereafter, Walter received an email from 0013's owner in Italy explaining that he wanted to buy the engine for his Spyder. Apparently, 0013 had raced at Sebring in 1955 with Jack's engine. As it happened, 0013 had a plain-bearing four-cam, so the owner asked if they could work out a trade. Well, Walter is no dummy. He knew what the original 0013 motor was worth, and he also knew that by having the original motor in that car it would increase the value of that Spyder by some $30,000. A motor trade was agreed with an additional $30,000 for Walter.

As Walter was boxing up the motor, Cabrera stopped by. "Jack, you can't just send off this motor to someone you don't know. We should really look this guy up." Some Google searching revealed exactly who this gentleman was. They discovered he was the organizer of the modern-day Mille Miglia as well as a former owner of the Dallara F1 team who possessed a Ferrari collection, including a GTO.

Curiosity satisfied, they shipped the engine, and in a few weeks, a shipping crate arrived. Inside was a four-cam engine freshly rebuilt by F1 mechanics. Of course, it would be nice to find another roller bearing crank engine to put in it, but the plain bearing engine is a better motor for regular driving.

Jack eventually turned this original Porsche Outlaw loose, and it now resides in Southern California. In 2023, the *Kustom Karrera* was part of the Porsche 75th Anniversary exhibit at The Petersen Automotive Museum.

Sadly, Dean Jeffries passed away in 2013 before Jack's restoration was complete. If he had lived long enough to see it driving through the streets of Los Angeles (L.A.) and onto the floor of the Petersen, he would have thought it was, to use his well-known phrase, "super bitchin'."

OPPOSITE, LEFT: Note the simple rear lines of the *Kustom Karrera* with bumpers removed and three features unique to this car: taillamps, motor intake grill, and roof vents.

OPPOSITE, RIGHT: There are two details on the *Kustom Karrera* that were inspired by the Mercedes 300 SL Gullwing. One is the rear roof vent and the other is the taillamps. It seems that Porsche wasn't the only German car Dean appreciated.

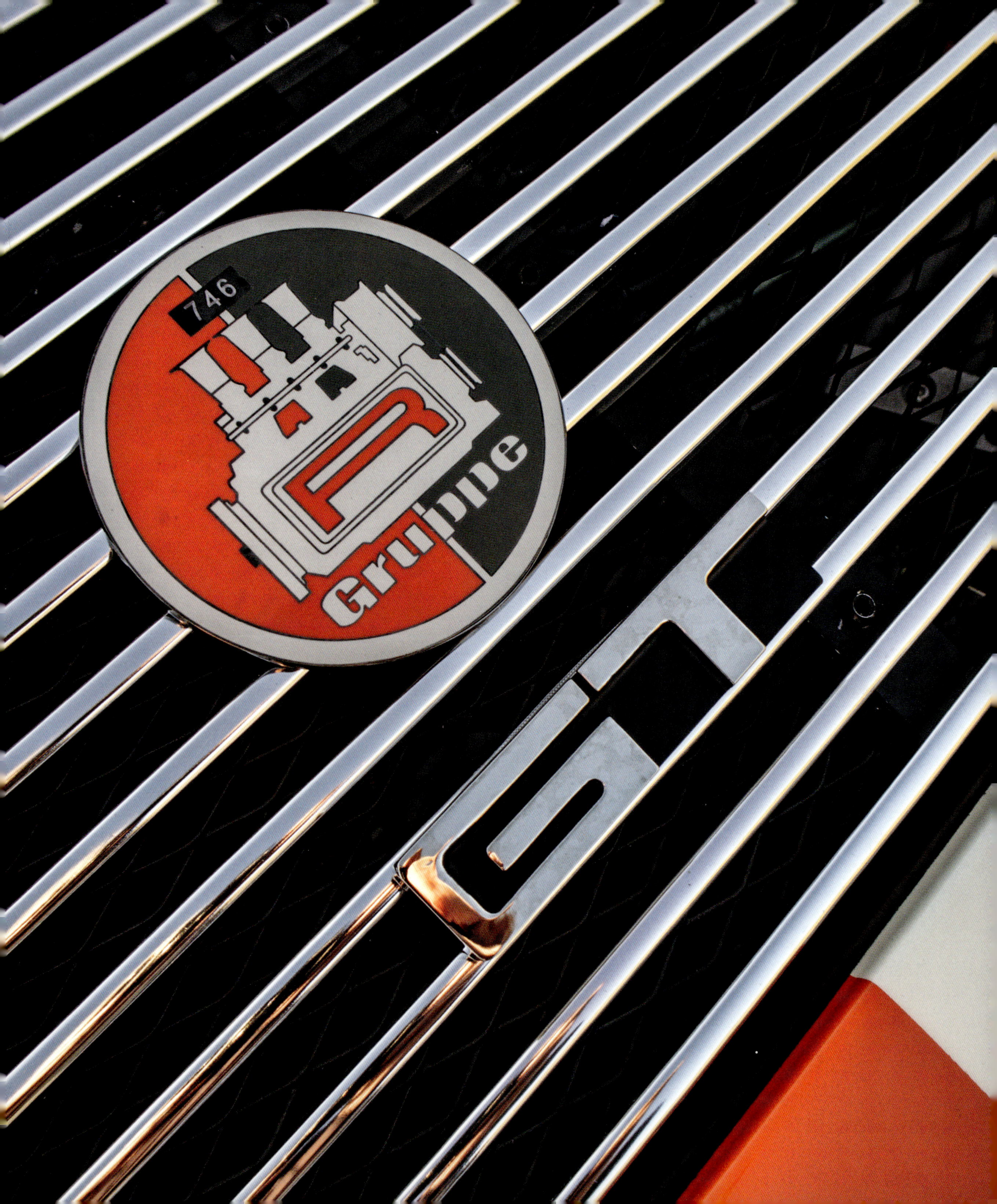
746
Gruppe
GT

R Gruppe badges are a complicated design. There have been several iterations though the years. This particular badge includes member number *746*. The badge at left is the highly coveted GT version presented to stand out cars and awarded at the annual R Gruppe Treffen.

THE R GRUPPE

A.K.A. "THE MERRY BAND OF MISFITS"

If the R Gruppe had a theme song, it would most certainly be "Don't Let Me Be Misunderstood" by the Animals.

Never in my entire life have I heard so many misconceptions about a car club. The comments range from "You don't want to mess with them" and "What a bunch of assholes" to things like "Stay away from them—they're a cult."

So, I sat down in person to discuss these assumptions with several club officers. I can tell you these criticisms could not be more of a mischaracterization.

TAKE IT EASY

To understand the R Gruppe, we need to go back to the early 1960s, long before the group even formed, to the now defunct Parts Central Salvage Yard in Berkeley, California. Parts Central was where a young Jim Breazeale worked dismantling Volkswagens. They were a cinch to deal with, coming apart easily, and you could sell all the bits as everything was in demand.

Along with the Volkswagens came the occasional Porsche. Breazeale was obsessed with all things Porsche, so he became "the Porsche Guy" at Parts Central. Any requests related to the marque automatically funneled to him.

Fast forward to 1978: Breazeale decided it was a good time to strike out and start his own Porsche-specific yard named European Auto Salvage Yard (EASY) just a few miles away in Emeryville.

Although Breazeale's former salvage boss wasn't always easy to work for, he was smart enough to offer that "we shouldn't compete with one another; instead, we should help one another." And help he did. He worked out a deal with Breazeale, whereby Jim took all of the Parts Central Porsche inventory to launch EASY and paid his boss back when he could at an agreed-upon price.

With a new business and start-up inventory, Breazeale hit the ground running.

A few years in, Breazeale noticed repeat Porsche customers showing up at EASY requesting a variety of suspension and performance parts for their cars with the intention of hot-rodding them. Enthusiasts modifying Volkswagens was nothing new, but Porsches? Of course, California speed freaks had a long history of hot-rodding going back to the 1930s, but still, Porsches?! Breazeale had a first-hand inkling of what was going on as he was driving a Porsche 912 he'd bought at auction some years before. It was a theft recovery that had lacked seats and an engine when purchased, and he had been hot-rodding it with salvage parts going back to his Parts Central days.

European Auto Salvage Yard (EASY) founder Jim Breazeale has more stories than you can imagine about Outlaw Porsches. Up until he sold it in 2017, you could find him brewing coffee on the first Saturday morning of every month for Cars and Coffee at EASY in Emeryville, CA.

As the years rolled by, requests for parts to upgrade engine, suspension, brakes, and body panels continued to grow. EASY became the Northern California (NorCal) source and hangout for like-minded Porsche hot-rodders.

EASY's legacy is reflected in Porsche Outlaws still on the road today. Richard Sutlett's yellow 1973 911 S and Joshy Robot's 911, with its signature red right door, are perfect examples. "That door came off the shelf at EASY," Breazeale recalls, along with Joshy's declaration that "I'm never gonna paint that door." Was Joshy cheap? Was it a statement? Who knows? Regardless, that door became a signature of the car and who Joshy is. Steve Hatch's stunning, badass 911 was pulled together with parts out of EASY. The list goes on and on.

EASY ran strong for 40 years until January 2018 when Breazeale closed the doors for the final time. In fact, if he hadn't encouraged his son, Richard, who grew up at EASY, to find his own path and become a commercial pilot, EASY would likely still be going strong.

NOT JUST NORCAL

This Porsche hot-rodding thing was not localized to NorCal and EASY's immediate locale. The same thing was happening in Southern California (SoCal) situated around a place called AASE Brothers, another Porsche parts supplier in Anaheim. And Porsche Dealers like Chick Iverson in Newport Beach and Vasek Polak in Hermosa Beach were inundated with requests for parts available through the *Porsche Sports Purpose Manuals*.

THE BIRTH OF R GRUPPE

Hot-rodding has many different definitions, as we know. From a Porsche perspective, it encompasses everyone's intentions—to compete on the track, to cruise the Pacific Coast Highway, or to rip up Mulholland Drive. Whatever made your car faster, cooler, or better handling was the name of the game. After all, movie stars and "real" car guys from James Dean to Steve McQueen and Paul Newman to Robert Redford all drove modified Porsches. After all, "cool is as cool does."

The hot rod Porsche trend continued to grow through the 1980s and 1990s, with younger customers showing up at EASY looking for parts for everything from 912s to 914s and even 944s. One of them was Pete Stout who was the Editor-in-Chief of *Excellence* magazine for Ross Periodicals. Stout assigned a story about Cris Huergas and a group of Porsche hot-rodders who hung out at EASY to Dave Coleman for the April 1998 Issue (issue #76).

After reading Coleman's *Excellence* article about Huergas and his NorCal hot rod Porsche crew, former Porsche Senior Designer Freeman Thomas (who was then with Audi) was struck by the observation that the craze of hot-rodding one's Porsche was not just a local SoCal thing, that it was something much bigger.

Thomas contacted Huergas, suggesting that they create a club together. After several conversations via phone, letters, and postcards between the two Porsche fans, the R Gruppe was established on May 15, 1999.

Those early communications were about selecting the club's name and sorting the group's focus, including the type of cars permitted as well as the style of hot-rodding that would define an R Gruppe Porsche.

You never know what will show up at EASY. Hot off the trails from the Baja 1000, this 924 S Baja car was happily welcomed by R Gruppe members in appreciation of its badassery.

WHAT'S IN A NAME?

Early club name suggestions included the following:

- 911 GS—Grand Sport
- 911 GTS—Grand Touring Sports
- 911 RSK—Rally Sports Kars
- 911 SL—Special Lightweights
- 911 LS—Lightweight Specials
- 911 R Gruppe—"R" Group

After several discussions, the name R Gruppe stuck. It was both a tribute to the Porsche 911 R and a play on the words *Our Group*.

To tie in the German roots, Group was spelled *Gruppe*—a two-cylinder word just like Porsche. Proper pronunciation is sometimes forgotten, but not by Thomas or other early members.

With the name sorted, the next step was to determine which cars would be accepted into the club. This generated another lengthy list of contenders.

ABOVE: Salvage yards everywhere are breeding grounds for Porsche Outlaws. The fodder to make your Porsche unique or keep it on the road is piled deep throughout.

LEFT: If you're willing to get your hands dirty, treasures abound in a pile of parts like this. Everything from suspension pieces to brakes and even that odd-sized wheel of which you need just one can make your day.

Here are just a few examples of cars that have been resurrected with parts from EASY. This was shot just days after EASY was cleared out and sold to a new owner. If those walls could talk!

R GRUPPE CARS

Initially, there were just three criteria for R Gruppe cars:

- 911 long-hood cars (1973 or older)
- Sports Purpose builds
- California hot rods

If sorting the name took considerable work, you can imagine the number of conversations, letters, and postcards it took between Cris and Freeman to narrow the eligible-car list. But the list makes sense on several levels from an inspiration standpoint.

Why long-hood cars? Go back and watch the opening scene of the 1971 film *Le Mans*. Steve McQueen's drive through the French countryside in that Slate Grey long-hood Porsche 911 S is the definition of cool. In fact, it's so cool it overshadows the fact that you are listening to a soundtrack featuring an oboe!

Jim Breazeale wasn't just the parts hustler to the R Gruppe. He's a member who *lives* the outlaw lifestyle. His Sepia Brown 911 can be easily spotted by its distinctive six tailpipes.

Sports Purpose Porsches are cars modified for maximum track performance while remaining street-legal. Thomas explained, "That's the Steve McQueen part of all of this. I'm not gonna show my hand, I'm gonna drive this [seemingly] docile little machine there, I'm gonna run it and drive it at the fastest time, and then drive it home."

Porsche had created its own manual, *The Sports Purpose Manual*, for those interested in turning their stock 911 into a Sports Purpose car. It was published in 1968 for the 911, 911 L, 911 T, and 911 S and again in 1972 for the 911 S and the 914/6.

These manuals became the bible for R Gruppe Sports Purpose cars.

The cool thing, back in the day, was that all the parts you needed were available directly from your local dealer's parts department. Porsche dealers were swamped with orders.

R Gruppe cofounder Thomas said, "The R Gruppe is an extension of no-ego, sharing creativity—an extensions of one's passion for the 911 and 356. It's all about the idea of Sports Purpose. That's what Steve McQueen was all about. When you consider the long history of hot-rodding you start to understand the R Gruppe. There was a time you could buy a used Porsche for $15,000 and not feel bad about cutting it up."

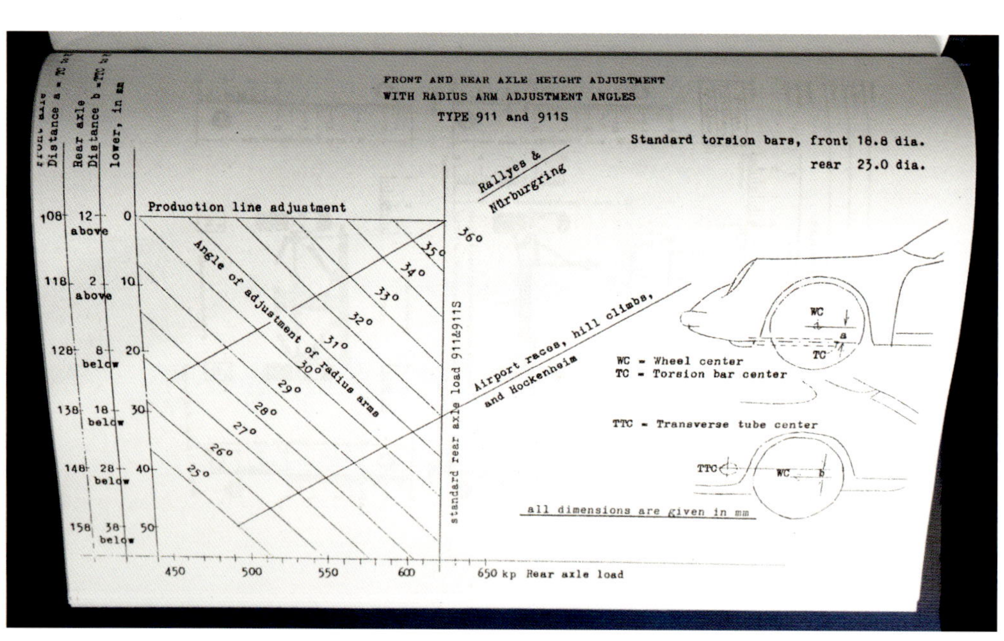

ABOVE: Joshy Robots' "Red Door" 911 has earned him a reputation for not giving a damn and backs up his motto of "Kill All Porsches" (translation: drive the wheels off of them).

RIGHT: Straight from of the pages of the *Porsche Sports Purpose Manual*, here are the front and rear axle height adjustments. The catalogue is the R Gruppe bible.

THE BADGE

So, R Gruppe had a name and a definition of what cars were acceptable. Now, it needed a logo and a grill badge not unlike the cloisonné badges European clubs sported.

Thomas' brother-in-law was a jeweler who also worked part time with Gary Emory, so he was tapped to help create a badge for the group. Several designs were considered, and the R Gruppe founders eventually settled on one that used a flat-six profile outline split by two colors: orange on the left and gray on the right. Placed in the carburetors of the original badge was *911* with a black frame around the numbers. Eventually, the club was contacted by Porsche Cars North America's legal department and asked to remove the *911* from its badge. Porsche owns 911 as its trademark and doesn't like it used anywhere else. Porsche was protective of their 911 marque after having to change the name of their newly debuted 1964 901 sports car to 911 due to a patent dispute with Peugeot.

The original badges had members' club number etched on the reverse side. When the badge was redesigned, it seemed logical to place the member's club number where the original *911* had been.

The badge has become a sacred thing within the club. If a member sells their car, they are asked to remove the badge because it represents the owner and not the car. The car is not a member of the R Gruppe. It's the *person* that is the member of the R Gruppe. That's a critical distinction.

The same holds true for any R Gruppe stickers that are on the car. There are official R Gruppe stickers, and those are supposed to be placed (just like any good hot-rodder's club sticker would be) in the inside lower right-hand corner of the windshield on the passenger side. Other stickers or badges from R Gruppe events can be left on the car, but any sticker or badge with a member number needs to be removed.

Like the handling of badges, the club has a few other strict rules, among them the expectation that a club member will never make money from the group name. For example, you can't offer your vehicle for sale as an R Gruppe car. Once again, it's the member that's part of the club, not the car.

This raises an interesting question: What sort of provenance is bestowed on a car previously belonging to an R Gruppe member? There are numerous examples of famous hot rods that are known by and named for the person that built that car. Is this the case for R Gruppe cars? It would seemingly be so. For example, founding member Cris Huergas passed away in 2020, and his car has changed hands a few times since, each time for more money. Many club members feel that the increasing value of the Huergas car is based on the fact that it was owned by the late cofounder of the R Gruppe, hardly an unprecedented situation in the hot rod world.

The R Gruppe badge has gone through several designs over the years. Some lack a member number on the front like this one (747). Those without a front-side member number have the number etched on the back of the badge.

The opening scene of Steve McQueen's *Le Mans* film features this Slate Grey 911 S as it meanders through small towns and the French countryside. Every car guy knows the scene: no words are spoken, just a lone 911 S on a beautiful drive to the world's most grueling endurance race. It's "the calm before the storm."
BFA / ALAMY STOCK PHOTO

MAKING THE GRADE

One sure way to *not* become a member of the R Gruppe is to ask how you can become a member. After being burned several times by folks whose interest in the club proved insincere, R Gruppe put in place a vetting process that can take up to a few years. As with any relationship, it takes time to figure out precisely what that relationship will be. One can't just show up at an event for the first time with a new build they finished last week and expect automatic membership.

So, what is the process? Begin by simply showing up. Attend as many R Gruppe events as you can with the group. Let them get to know you. When invited to join them for a drive or a cup of coffee, GO! When it comes to being a possible member, it's all about who you are. The club is more concerned with who you are than what your car is about. The car is merely a calling card, and a member's car may change over the years. What would you do if someone breaks down on the road? Would you honk and wave as you go by, or are you the person who will pull over, get dirty, and help out? That's what the club brotherhood is all about: knowing that fellow members have your back at all times.

I've always believed that a man's true wealth is measured by the number of people he can call at 2 a.m. asking for help and knowing those on the receiving end will show up no questions asked. Before SoCal Chapter Meister Johnny Riz (short for Risvold) was asked to join, Cris asked him—knowing Johnny had a trailer and a truck—that if an R Gruppe member broke down on the road in the middle of the night could they count on him to show up and help out? Johnny's answer was an emphatic YES.

NorCal Chapter Meister Koichiro Kanda, a.k.a. Koich, echoes this member ethos:

> **"It's about who you really are. More than anything it's a brotherhood. Member decisions are made by asking yourself 'do [I] want to be sitting across from this guy when [I'm] having dinner?' Or do you want to go on drives with him? Do you trust him behind the wheel at high speeds? We look at a person's character and decide whether or not [they're] a fit. If they might be a good person? Could they be a wild card? Of course, wild cards also can be a great contribution to the club just to keep the edge. There is a vast variety of characters within the group. Some are really wild, and some are docile, just like any family. Of course, if you're too much of a wild card you get booted. You can never be disrespectful to other members or cause problems which would put the club in a tough place."**

Once you put in your time with R Gruppe and members get to know who you are, then eventually, if you're lucky, a member will represent you and suggest you to a Chapter Meister. The Chapter Meister will then review who you are, what you're all about, and discuss the possibility of you becoming a member with the rest of the chapter. Once that decision is made, the actual moment of becoming a member is rather unceremonious. There's no fanfare, no fireworks, and no complicated initiation rights. You're simply in. Club member Jeff Higgins explained, "We're kind of like The Borg [in *Star Trek*]. You just get absorbed. One day, two or three years down the road, someone walks up you, shakes your hand, hands you a sticker, and welcomes you to the club."

BELOW LEFT: NorCal Chapter Meister Koichiro Kanda with his Porsche Outlaw that he handbuilt in his home garage is the real deal. There isn't a nut or bolt of this car that he doesn't know. With the help of friends and the patience of his wife, this G-Body Outlaw came to life.

BELOW RIGHT: In 2019, R Gruppe celebrated its 20-year Anniversary. This is one of the commemorative grill badges that included the words *Sports Purpose* on it.

(continued on page 64)

GLORIA RISVOLD

Gloria Risvold is one of only five women to become an R Gruppe member since its inception. Her husband Johnny "Riz" is also a member, but Gloria's love of Porsches goes back even further than Johnny's.

AT 19, IN THE MID-1970S, Gloria decided college wasn't for her. Her first job out of high school was at K & E, a Porsche repair and restoration shop in Costa Mesa, California, known at the time for dropping American V-8s into 911s. Gloria adored being surrounded by Porsches every day and credits shop owner Richard Eaton for igniting her passion for the brand.

After several years of DMV and clerical work in restoration shops, Gloria ended up at Dot Datsun in Huntington Beach, California, where she mastered the art of California DMV paperwork. The store processed 150 used cars a month, and Gloria became the go-to guru for every deal. She was making a name for herself as her DMV skills were invaluable.

Her next move was to Pristine Motorsports in Huntington Beach, California, a dealership she would call home for the next 39 years. Pristine is well known for its custom coachwork and auto sales, and though they deal in other German brands such as Mercedes-Benz and BMW, they are mostly recognized as a pre-owned Porsche shop. It became common practice at Pristine to regularly toss the keys to Gloria for a car that needed a post-service shakedown run or similar exercising, affording her the opportunity to get behind the wheel of some very cool Porsches.

One day in 1997, she brought home a Guards Red 1973 911 S that never made it back to the showroom floor. She and Johnny fell in love with the car, and they are its proud owners to this day. The Risvolds eventually hot-rodded the S, inspiring them to join several Porsche clubs in a search of a fit. One day, Johnny spotted an article in *Excellence* magazine about like-minded Porsche owners. The R Gruppe looked like a place they could call home.

Gloria's love and involvement with the Porsche community grew deeper. She became instrumental in the planning of the R Gruppe's annual gatherings and group meetings. She became a regular on drives like "Pines to Palms," and when Johnny's car was in the shop, they would drive one of her 911s, like her 1972 Metallic Gold Targa or her favorite, a Grand Prix White 1983 Euro spec 911 Turbo. As Gloria notes, "That Targa is way too slow for me."

After spending decades immersed in the R Gruppe world, Gloria still didn't see this coming: One Saturday night, during an R Gruppe meeting's evening announcements, Cris Huergas called her up on stage and presented her with her own member number, 658. She became one of only five women in the club, not because she was a woman, but because she is a hard-core Porsche enthusiast.

OPPOSITE, TOP: SoCal R Gruppe Chapter Meister Johnny Riz's car gleaming in Guards Red at the Newport Beach Yacht Club with Balboa Island in the background. Riz's passion for driving and sailing are legendary.

OPPOSITE, BOTTOM: Gloria Rizvold is one of only five women ever invited to be an R Gruppe member. Rizvold's knowledge of the cars and genuine enthusiasm for club events made the membership decision a no-brainer for Cris Huergas.

Window stickers are an art form and represent the sentiments and pride of the owner. Note the R Gruppe 20th Anniversary sticker upper left and the R Gruppe sticker in the lower left with *911* located over intake velocity stacks.

(continued from page 61)

HOW MANY MEMBERS?

That's a complicated question to answer.

The first group is composed of the founding members. This is a small cadre who in the beginning helped create the R Gruppe. These are people like Freeman Thomas, Cris Huergas, Rolly Resos, Ernie Wilbert, and Jeff Zwart.

Once the R Gruppe was established, it was agreed that membership would be limited to no more than 300. This holds true to this day for what is known as *dues-paying members*. If you're a dues-paying member, you're obligated to a $50 per month fee to help support the costs.

There are also honorary members who represent the spirit of the group, for example Hurley Haywood, Jim Breazeale, Bruce Canepa, and others.

As previously discussed, each member has a unique number. If that member leaves the group or passes away, their number is retired. In this way, the numbers themselves are ever incrementing even though the R Gruppe holds at 300 issued numbers. Lower numbers are prized. For example, Freeman Thomas is #003. Jeff Zwart is #011. Members can also request specific numbers, like Bruce Canepa's number 962.

In more than 20 years of existence, R Gruppe has expanded into a global organization with members in Europe, United Kingdom, Netherlands, Germany, Spain, the Philippines, Singapore, and Australia.

YOU GOTTA DRIVE THEM!

In the club's early days, members would gather on routes along the way to the annual *Treffen* (meet) to enjoy a "spirited drive" together.

In time that became frowned upon, a sort of "we don't do that anymore" situation. It was deemed too reckless and not what the R Gruppe wanted to be about.

But instead of canning the spirited driving entirely, R Gruppe did the "mature" thing and decided to gather at a racetrack on the way to the annual R Gruppe Treffen. Here, they drive hard and fast on a controlled course they have rented for the day. Johnny Riz describes the track time as "shaking a soda can until there is no more fizz left." The need for speed sated, everyone cruises to the Treffen location to enjoy a few days of amazing drives in the gorgeous countryside at a more discrete pace.

Jeff Higgins from the PacWest division recalled one of those memorable drives.

> **"I remember one day we got blocked by the Idaho State Police at a rest stop. They blocked the entrances on both ends. Then one of the officers stepped up to me and said, 'I see you guys are driving Porsches. We're getting lots of reports and mobile phone calls about a group of guys driving Porsches in an unsafe manner and just being jackasses about it. Seeing as how you guys are driving Porsches maybe you're going to meet up with that other group that's doing these things and maybe you can pass a message on for me that we don't appreciate that and to knock it off!'**
>
> **"Coolest guy ever! He finished with 'oh by the way, the Oregon border is forty-five minutes that way. Don't be early.'"**

This is a great illustration of the well-honed racing tradition of losing weight wherever you can. If you're going to run a metal shifter knob, you better punch it full of holes.

This is a perfect example of what following the *Porsche Sports Purpose Manual* can offer you in terms of stance. Not only will the look be spot-on, but the handling will be drastically enhanced.

R GRUPPE CHALLENGES

The R Gruppe faces some interesting challenges in the road ahead. Given the ever-increasing cost of a Porsche, donor vehicles can be expensive. Even 10 years ago, you didn't feel bad about cutting up an inexpensive car. Today, there are virtually *no* inexpensive cars.

Many of the members today are ageing just like their cars. As time goes on, R Gruppe President Bob Aines suggests that "You might start seeing things like a 964 RS [though not in America] being accepted in R Gruppe. That's a Sports Purpose car! Of course, the guy still has to be a good guy. Other cars which may eventually be recognized would be an SC RS, 1973 RSR, and more factory Sports Purpose cars. There are G-body cars in the club [now], and I think that will be the evolution. I have a 1969 911, and I have driven that to California fifteen times and to the East Coast ten times, and I'm tired of driving that car long distance." That presents a challenge based on the fact you are expected to *drive* your car to the Treffen.

Johnny Riz from the SoCal chapter points out how expensive the long-hood cars have become, which also suggests that the range of acceptable vehicles will have to expand. He also stated, "I'm really against the segment of people who just want to be associated with the R Gruppe, and I don't want the club to become a yacht club. It's getting harder to read through the posers to find the real deal these days."

ABOVE LEFT: Safety is always a factor to be addressed when increasing a car's performance. The roll bar in Steve Hatch's car was created and installed artfully.

ABOVE RIGHT: "Increase lightness" is everywhere you look. Even Steve's door handles are drilled to reduce weight.

The contrast between founding members and new members is best summed up by Johnny Riz: "The first few *Treffens* . . . everyone had a red shop rag hanging out of the back pocket of their jeans and dirty fingernails. These were cheap cars back then, [and] we were guys who could afford the car if you knew how to work on it."

Jeff Higgins from the Pacific West (PacWest) group noted, "We have already seen guys pull up in a diesel pick-up, park down the street with a trailer, offload their vehicle, and drive two block to Treffen. Talk about a party foul!

"One thing that will not change when it comes to R Gruppe cars is that they must be of the air-cooled era. The other factor is the cars have to be driven. We're not really looking for members that have a great garage full of cars. We look for members who enjoy their cars."

NOT WHAT THEY SEEM

We've all become familiar with "the velvet rope"—that barrier that bars someone from entering the coolest club in town or that wristband that grants you access to a higher level experience at an event.

After spending time with the R Gruppe, I've determined that there is no velvet rope around this group. That perception is fostered by people that aren't affiliated with the group and may not even know its members. These are just down-to-earth Porsche guys who love to drive and mod their cars.

Jeff Higgins said it best, "We're just a bunch of doofusses with cool cars."

Every car has a story. Thanks to imagination and creativity, this car's story continues in a rapid way. Driven hard by owner Steve Hatch, it shows no signs of being parked any time soon.

You would never guess by looking at this car that it has rolled over 400,000 miles (643,378 km) and was driven on all seven continents before being transformed into the R Gruppe car it is today.

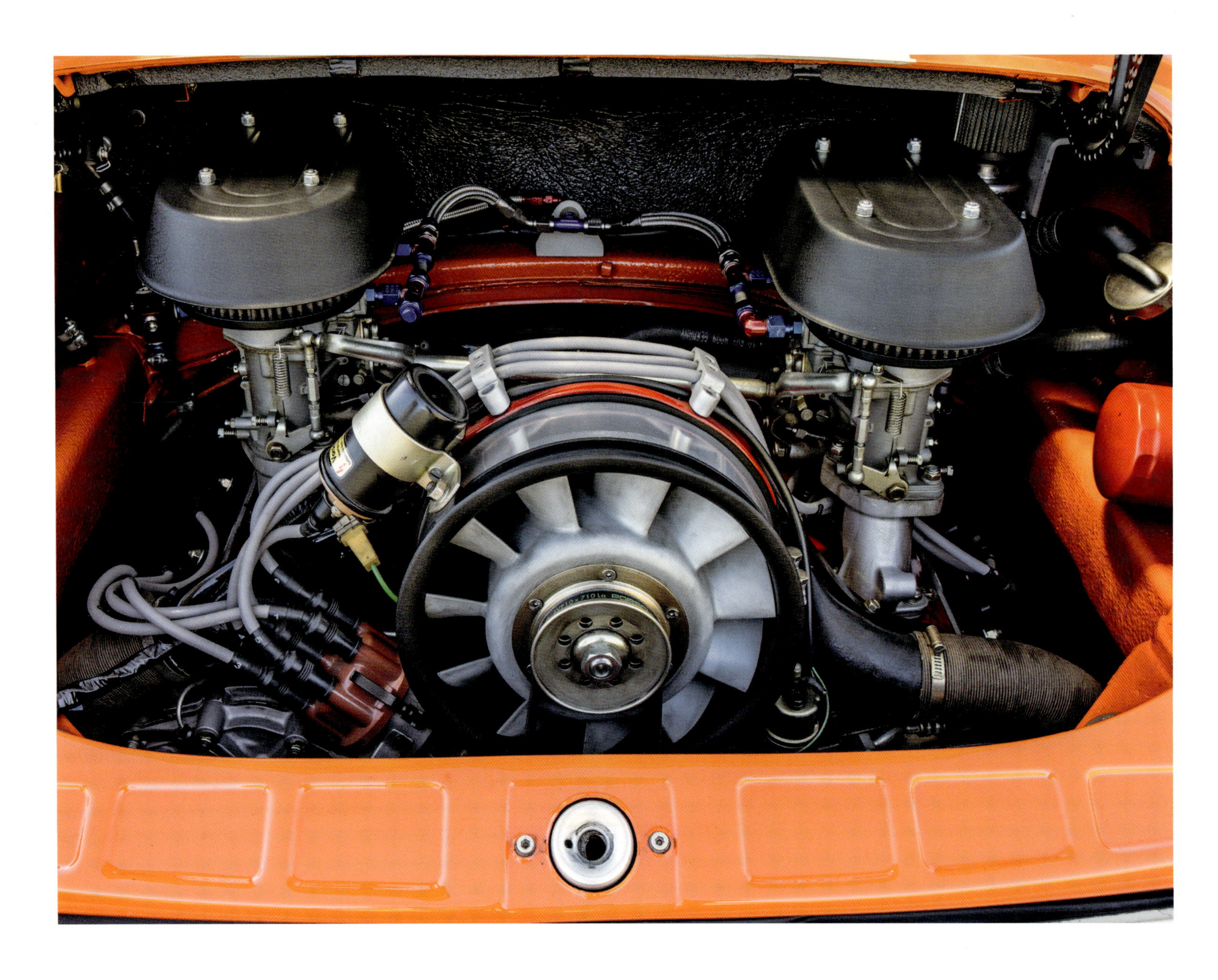

ABOVE: Lurking beneath the hood of Steve Hatch's 911 is a 3.0-liter engine with 401DA Weber carbs and a mild cam. It not only looks good—it screams!

OPPOSITE: Vic Rola's 1972 911 E Outlaw has long been a favorite. It's considered one of the finest examples of a true R Gruppe car. Don't let the shiny paint fool you . . . Vic is more than happy kick up a dust storm given the opportunity.

356
OUTLAWS

5

THE BUILDERS

ABOVE: Perhaps the finest example of the apple not falling far from the tree. Gary Emory (left) and his son Rod Emory (right). The true "Outlaws."

TOP: Somewhere in North Hollywood . . . in a nondescript building branded by a single German word *Bäckerei* (Bakery), Rod Emory and his family create masterpiece Porsche Outlaws.

EMORY MOTORSPORTS

Let's see if we can get this story straight. To do so, we need to go back to the 1940s and Rod Emory's grandfather, Neil Emory, who owned Valley Custom Shop.

Neil grew up in Burbank, California, which was and still remains, arguably, the center of US car culture. This location was in close proximity to the movie industry and its discretionary income and home to shops like John von Neumann's Competition Motors and the So-Cal Speed Shop.

When Neil was a kid, he would cruise the Warner Brothers dirt parking lot, leaving little tags on the windows of the cars offering to clean them or gas them up. His excellent work and reliability built trust in the community. Soon, people were hiring him to perform all kinds of work on their cars.

In World War II, Neil served in the Navy. He thought he was going to be shipped overseas, but instead, he was assigned to the motor pool in Alameda, California. Late one night, an officer crashed his car and needed it fixed fast. Neil hadn't done a lot of bodywork, but he was able to bang out the fender and make it look presentable. In short order, he was running the motor pool and never did get shipped overseas.

When his service was complete at war's end, Neil teamed up with Rod's great uncle Clayton and opened Valley Custom Shop in 1948. Their first customer was Alex Xydias' So-Cal Speed Shop. Valley did metal work for the famed So-Cal Special belly tanker and eventually ended up building the equally famous aluminum So-Cal Streamliner. The Streamliner was the first hot rod to exceed 200 miles per hour (322 kph).

This is the Emory clan, including the latest addition, Drew Hafner, who married Jayde Emory in November of 2022. (L-R: Drew Hafner, Jayde Hefner, Zane Emory, Amy Emory, and Rod Emory)

Neil wasn't known for simply chopping cars. He took a more holistic approach, considering the car in its entirety and determining how he could refine it. Valley was known for applying elements of European styling and having fun with the modifications. Neil would never do anything so drastic that it stood out from the rest of the car. The list of famous hot rods that rolled out of Valley Custom Shop between 1947 and 1961 is an honor roll of hot-rodding. Ron Dunn's 1950 Ford coupe, Ralph Gill's 1940 Ford, Ina Mae Overman's 1952 Lincoln Capri, and Jack Stewart's 1950 Oldsmobile Holiday 88, a.k.a. the Polynesian, are just a few. All of these cars reflected the Valley philosophy of moving a vehicle's lines by channeling and sectioning while accomplishing it in a way that demanded study to figure out what had been done. In Neil's words, "Don't half ass it." The key is to modify everything, but make it look like it wasn't modified.

Valley Custom Shop closed its doors for the last time in 1961. Neil Emory went to work for Chick Iverson Porsche in Newport Beach, California, running its body shop.

When Neil's son Gary (Rod's father) graduated high school, he went to work at Iverson as well, eventually becoming the parts manager. This is how Rod Emory would become connected with Porsche. "Otherwise, I'd probably be sectioning and channeling American hot rods," Rod notes.

While Neil and Gary were working for Chick Iverson, Gary—who Rod describes as more of a salesman than a visionary—came up with a pretty brilliant idea that he presented to Neil. Working between the back workshop at Chick Iverson's dealership and Gary's home backyard, they built the first Baja Bug, which went on to become a California craze.

Being inside the metal shop at Emory Motorsports requires full-time ear protection, which was graciously provided by my host. It's nonstop serious business here as the crew literally pounds out the next masterpiece.

Rod Emory was born into this multi-generational gearhead scene in 1974. He remembers childhood days spent with his father at Iverson's digging through the parts department where he learned all about automotive components. "I knew my dad's parts shop better than he did—from 3 feet [1 m] down."

"My father loves to tell the story about a customer who came in asking for a cam fitting or oil fitting. My dad told him, 'I don't have them.' I think I was five or six years old, and I said 'I'll get you one.'"

Rod knew exactly where the bin was. He spun around, retrieved it, and placed it on the counter. Gary looked at him and said, "I'll never question you again."

Rod was about six years old when his father and Iverson started Parts Obsolete in Costa Mesa, California. It was a 10,000 square foot (929 square meter) warehouse filled with a collection of vintage Porsche parts that Porsche couldn't or didn't want to deal with. The inventory encompassed everything from 356s to early 911s to racing 904s, 906s, and 908s.

After Rod's grandfather retired, he moved to Fallbrook, California, where young Rod spent his summers and learned firsthand how to gas weld and work metal. In between heating and hammering a metal project, Neil would hand the torch to young Rod to hold. One day, he turned to Rod and instructed him to heat the spot he was working on and then hammer it out. The die was cast. Rod recalls

Cars being assembled are surrounded by those waiting their turn. In the far right corner of the shop waits one of the last unfinished cars from Valley Custom Shop.

the times in Fallbrook as some of the best of his life. The skills he learned from Neil are employed every day in his own shop. Learning from one of the greatest hot rod builders in the world also instilled the confidence that has helped make him who he is today.

Rod was just 12 years old when he started working for Tom Topping, who started the Hi-Performance Swap Meet in Long Beach, California, and also owned a specialty fastener company. Rod worked for Topping at the swap meet driving around in a little Cushman cart and collecting money from the vendors. One day, Topping turned to him and said, "Hey Rod, do you wanna work as a drag steward?" Rod had a one-word answer: absolutely.

Amazingly, for the next three years, until he was 15, Rod traveled the country with Tom Topping and his racing crew. He was the left-side mechanic of a five-"man" team of boys between 12 and 15. They could fully strip and rebuild a hot top-fuel motor on a stand in 45 minutes between runs.

Rod left drag racing in 1988. About that same time, Gary had found a decrepit 1953 Porsche 356 coupe that had been rotting in a barn for years. Gary gave it to Rod on one condition: He would supply all the parts if Rod put in all the work to refurbish it. This would become Rod's first Porsche. It took him two years to complete the resurrection while working alongside his dad at Parts Obsolete.

Rod went on to help his father build another five or six cars in a space at the back of Parts Obsolete. Rod worked in a little bay in the rear of the shop following up on assignments from Gary like, as Rod recalls, "Go pull the hubcaps off of this; go drop the suspension; go put hood straps on this car, and install some fog lamps on the hood of that car." Gary knew Rod had the ability, so he threw all kinds of stuff at him. It was there at the back of Parts Obsolete that the "outlaws" concept germinated.

While Rod worked in the back, Gary was up front selling new old stock (NOS) Porsche parts destined for full concours restorations. Concours in the 1980s was already over the top with owners completely over-restoring their cars. "Cars weren't even that good originally," Rod says.

Drew Hafner hard at work creating yet another Emory Motorsports masterpiece. Every inch of the car body will bear his signature by the time he's done.

When all is said and done, you'll not *see* every detail that went into creating the finished product of anything coming out of Emory Motorsports . . . but you'll *feel* it.

"So imagine, if you will, in the front of the shop everything is about NOS and full concours quality. Meanwhile in the back of the shop, I'm out there dropping cars and putting fog lights on the hoods." Gary's customers thought it was blasphemy, complaining that you can't cut cars up like that. He responded that "We can kinda do whatever we want. If you guys think we're ruining cars, I guess we're just gonna be *outlaws* and that's that."

Around this same time (somewhere between 1987 and 1989), Freeman Thomas appealed to his brother-in-law, James Cannon, (a jeweler in Tustin, California) to create a badge for the Emorys combining the word *outlaw* with the graphics from a German five Deutsche Mark coin. (Gary had given Rod a 5 DM coin when he was three years old, and it's been around his neck ever since.)

Gary and Rod put the "outlaw badge" on a couple of cars that they had built, and Rod put one on the back of his car. "It was just for fun," Rod recalled. "We didn't start a club. We didn't sell the badge. It was just something we put on cars that we were building."

When Rod was 16 years old, he got his racing license from Jim Russell Racing Driver School at Willow Springs and started racing with the Classic Sports Racing Group (CSRG) when he graduated high school in 1992. That same year, he was at the Portland historic races when Cameron Healy (whose Kettle Potato Chips sponsored the event) approached and offered him some chips. He also said, "I would love to have you do what you did with your car to one of mine. It's beautiful." The two became fast friends, and Rod built a 356 cabriolet race car for him. He then offered him track support and started teaching him how to be a better race car driver. "I kind of fell into the business of building cars for people because of my friendship with Cameron."

A body in bare metal reveals the craftsmanship and true DNA of an Emory Outlaw.

In 1998, Porsche celebrated its 50th anniversary and all eyes were on the Monterey Historic Automobile Races. Rod and wife Amy had been married for about a year, and Rod decided to build Amy a Porsche. She loved the idea of a convertible. A few months into the build Amy realized she was pregnant, but they decided to move ahead with the project anyway. This was not going to be one of Rod's typical 356 builds or an outlaw. It was what would become Rod's first Emory Special.

Every Emory Special is an example of the lineage of creativity handed down from his grandfather, Neil. Every body panel on the car is reshaped or altered in some way. When you look at an Emory Special, you'll see several cues taken from the 550 Spyder. The sides of the car are rolled in, the front nose slants back, the louvered rear deck lid is larger, and the rear valance is completely different, including the frenched taillamps.

After debuting Amy's car along with an aluminum Eriba trailer they were towing in Monterey in 1998, the car then appeared on several magazine covers and in articles throughout the Porsche community. The car was a hit! Just about everyone Rod was representing with his vintage race service jumped in line to have him build a unique car for them.

Rod's hot rod Porsche building business started growing along with his thriving race support business from 1996 until 2008 when everything got a little strange. The race support business was averaging 12 to 15 events per year, with 30 clients, as well as building one or two specials a year and one or two Porsche Outlaws per year. In the fall of 2008, Rod could see things changing and decided if he didn't have a 40 percent commitment from his racing clients for the next year, it was time to do something drastic. So, he put out a questionnaire to his race clients asking for their commitment level going forward. As it turned out, he only received a 35 percent commitment confirmation from his race clients for 2009. With that information, Rod turned to Amy and said, "We're selling the trucks." He proceeded to sell their two semi-trucks and six car transporter. They stopped vintage race support altogether and have put all their efforts into building cars ever since. Talk about turning lemons into lemonade.

Emory Motorsports now produces 12 to 15 cars a year out of his shop in North Hollywood. If you like what he does, you'll have to get in line. His lineup of cars range from a 356 Outlaw to an Emory Special and now the Emory RS, which is a radical 356 grafted to a 964 chassis with all-wheel drive. Rod's attention to detail is off the charts and everything about his cars stem from that one simple phrase from his grandfather, "Don't half ass it."

LEFT AND NEXT PAGE: The sheer work that goes into an Emory Special is mind-boggling. Just like the best hot-rodders before him, Rod makes it all look like it was meant to be that way. The car you see in the build process here was unveiled at Luftgekühlt 9 in 2023 and pictured on the next two pages.

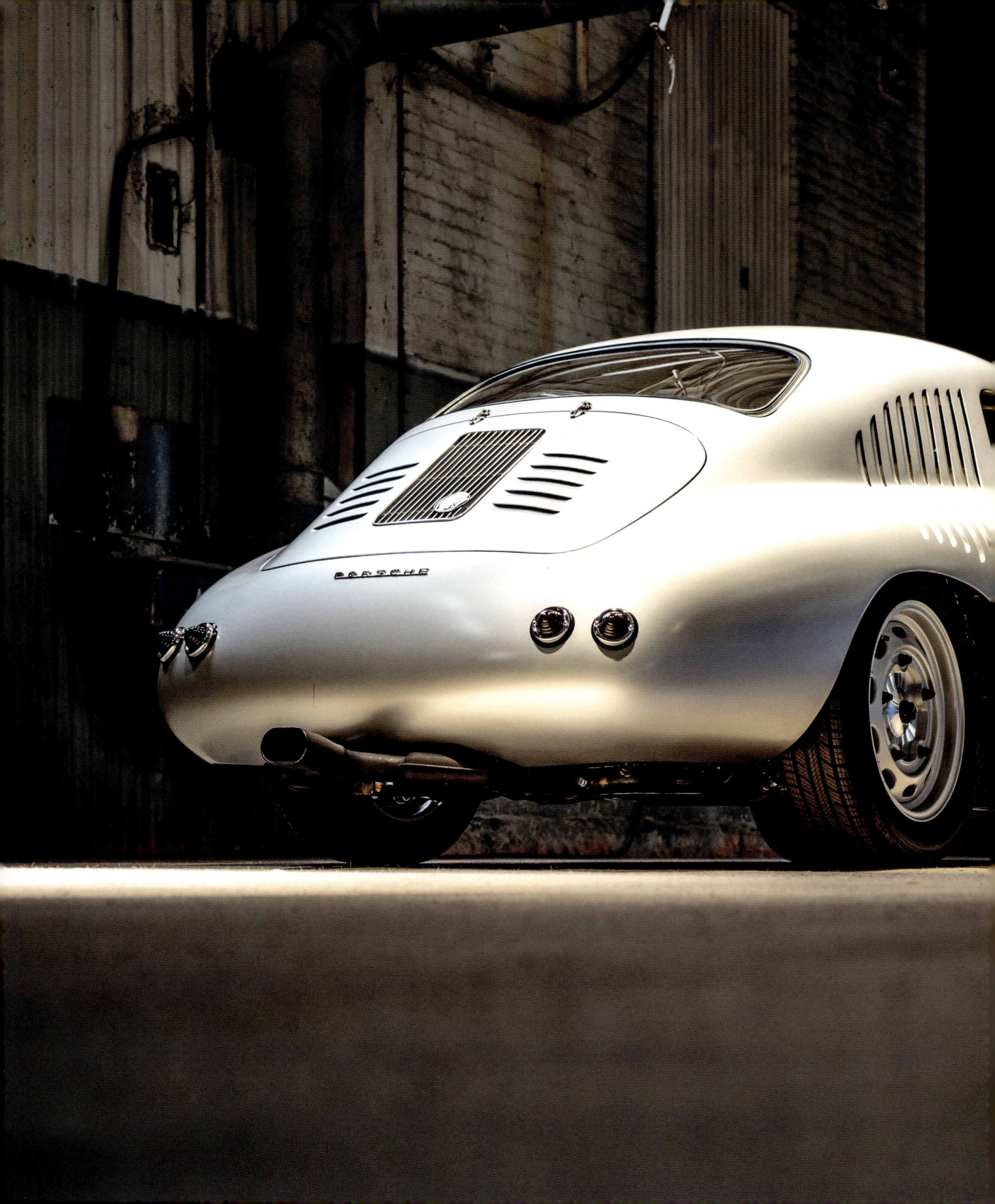
PORSCHE

BISIMOTO

Bisi Ezerioha has one of the most interesting back stories you'll find in the car business today. He was born in Nigeria in West Africa, the son of an accomplished geologist and a brilliant biochemist who both recall Bisi's first word as *car* and second word as *light*. Those early utterances make a lot of sense because Bisi is all about building cars that are powered by what you might call lightning.

As a child, Bisi was known for his curiosity. He couldn't seem to leave anything alone. If he had a radio, he wanted to take it apart and put it back together. Life was all about understanding things and how they work. Bisi's parents encouraged his curiosity and would help him take things apart and put them back together. But they also offered discipline for understanding. "I had to do a few things to make them happy. Of course, I wanted to please my parents—they are the Gods I could see."

Bisi's parents saw cars as just a mode of transportation. But Bisi became obsessed with cars. He found racing appealing, and he learned about cars through magazines like *Car and Driver*. "I also loved Japanese cars," he recalled. "They had some cool trends which I found really interesting." He also fell in love with Pontiac's Fiero. "Such a great platform, and so cool looking."

Bisi first drove in his father's compound in a Peugeot 504, a vehicle absolutely nothing like the cars he was attracted to.

After years of curiosity and his will to please his parents with good grades, Bisi decided to prepare himself for higher education. At age 14, he took the entrance exam for college so that he could practice over the next several years and be ready when the time came. Much to his and his parent's surprise, Bisi passed the exam to become a Petrochemical Engineer on his first try at age 15, becoming the youngest Nigerian to ever enter college.

Inside the Bisimoto shop, we find a perfect example of before and after. What a transformation!

After entering college, he became frustrated with what was available to him in West Africa. He wanted better learning tools, not old textbooks from the 1960s. He discussed his frustration with his parents, both of whom had been educated in the United States, and worked out a deal to attend school in Los Angeles (Cal Poly Pomona and Cal State University, Long Beach).

Bisi's mom escorted him to the international airport in Lagos, Nigeria, and many hours later, he landed at LAX with two suitcases and looking for a family holding a sign that said simply "Bisi."

Bisi's first car in the United States was a Honda CRX. It was a base model, which offered great gas mileage, and he had no intentions of modifying the car in any way. Then one day, he noticed the car was making a funny sound. He brought the car to the dealership, and they wanted $350.00 to repair a puncture in his muffler. Instead, he took the CRX to a private shop that installed an aftermarket muffler called the Dynomax Ultra Flo.

"Believe it or not, that changed my life!" Bisi said.

That simple modification immediately changed the car. First off, it sounded cool. And it seemed faster and better to drive. He also noticed the car got better fuel economy.

"I thought to myself, 'Wait a minute. Something's going on here. What is going on?'" As usual, Bisi's curiosity kicked in, and he learned all about back pressure, engines, and how science applies to internal combustion.

More thoughts spun up, "So, I just remove the back pressure and now the engine is better? What else can I remove? What else is placing restrictions on the car? What can make it go faster?"

His curiosity about making things faster brought Bisi into the "interesting" world of illegal street racing. He began to apply everything he had learned towards his CRX. Bisi learned a lot from books and YouTube, but when he asked people how they did things, no one wanted to share. The common response was "Just pay me and I'll do it for you."

"No one would explain why a car was fast to me," he lamented. "This is one of the reasons I [now] have my Tech Tuesday live broadcast every Tuesday on Instagram. It's something I wish I had when I was younger."

After racing with normally aspirated cars, Bisi started integrating turbos and became quite successful. Then one day, one of his clients said, "Bisi, you need to come to the Porsche world, you would really shake things up."

By 2008, Bisi had completed several client builds for shows like Specialty Equipment Market Association (SEMA) and found himself in a position to buy his first Porsche, a 1976 911. It was only a rolling shell, but it was his car. To shake things up, he installed the motor out of an early 996, but not before addressing issues common to the 996 engine like the IMS bearing. He then added massive twin turbos. The first time he drove the car in his complex, he almost killed himself. "It shut down my Dyno. I never had four digits ever in my life except for that [1976] Porsche. It was just absolutely ridiculous. That turbo would lag, lag, lag and then come on like a light switch!"

As Bisi's career progressed, he started building cars from major original equipment manufacturers (OEMs) for auto shows in Los Angeles, Detroit, Chicago, as well as the annual SEMA show. "After a few years, I noticed something: every year

BISIMOTO

TOP: Simple three-tone graphics nose to tail accent the lines of the extended Kramer body. The stark lines bring out the shape, which might otherwise be lost in the flat-black paint.

LEFT: A modern twist on the signature lower valance headlamp placement of a Kramer body includes white horizontal marker lamps.

OPPOSITE: What better way to draw attention to your work than by being as bold as you can be? Meet Bisimoto's Moby X. The decision to start with a Kramer 935 body as an EV platform was a brilliant move. Brixton Forged deep-dish polished wheels really offset the flat black paint job.

instead of my build budget going up [from the car companies], my budget was going down. Then I was asked to do a car for Harman Kardon for the Consumer Electronics Show [CES] in Las Vegas. I built a crazy, blue Hyundai Tucson slammed on bags, 700 horsepower [515 kW], big shiny Thurow [supercharger] coming out of the hood. It was amazing. Harman Kardon then provided all the audio equipment. It was an amazing sound system that made the inside of the car sound like you were in a theater."

When Bisi went to the convention center to drop off the car, he noticed something. There were cars from Toyota, Ford, Mitsubishi, and all these companies for whom he had built SEMA cars. They all had booths at the CES, and it was obvious that the budgets for that show were much bigger than for a standard auto show. He knew then and there it was time to stop being reluctant about embracing the new technology of electric cars and present himself as an outfit that could build similar cars for them.

In 2011, Bisi attended the Porsche Rennsport Reunion at Mazda Raceway Laguna Seca and fell in love with a Porsche 935 he saw on the track. "I just saw this fire-breathing twin turbo; I'm like 'oh my God.' I watched every video." He met several people that weekend, thanks to his outgoing personality, among them a gentleman who was curating some of the 935s. Bisi asked, "What do you do for body panels when a car like this gets damaged?" The answer was "It's simple. We made forms for the body panels, and we can just bang out new ones."

When one of his show clients came to him for a project, Bisi suggested building a 935 Kremer as the feature car. No one had ever done that before. After a brief conversation and handshake, Bisi went to work. He was so excited that he broke his own personal rule and began investing in the project with his own money before the deposit came in. Unfortunately, the project was scrapped and Bisi was left with a car shell and no client.

BOTTOM LEFT: Not the usual Porsche instrument cluster, but what else would you expect from a car of the future? Check the Bisimoto Engineering logo above the "P" as in Park.

BOTTOM RIGHT: This is not your typical frunk configuration. The Moby X has something different everywhere you look. You have to love the use of Porsche gas caps covering charging ports for level 1 and 2 charging via J1772 plugs.

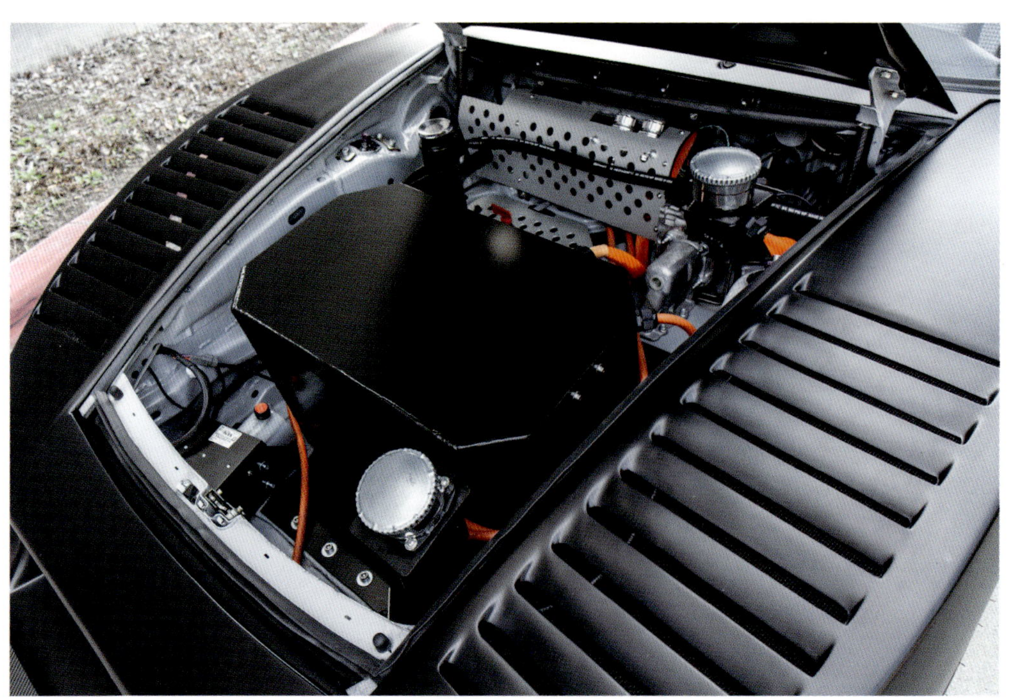

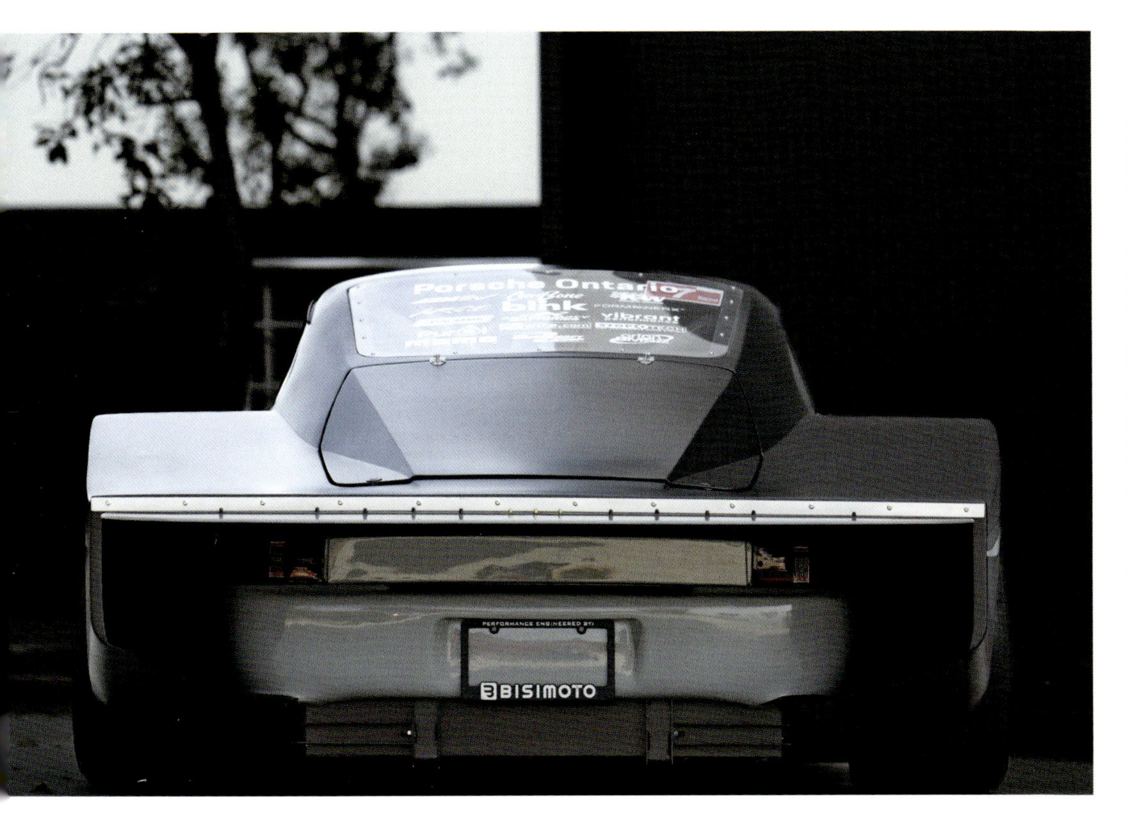

Bisi needed to make a decision. Rather than backing out of the aborted 935, he decided it was the perfect platform to embrace electric vehicle (EV) power.

In a reprisal of his early internal combustion days, he found himself knocking on doors, making phone calls, spending hours on the web, and trying to figure out how to make this technology work. He grew tired of people saying, "Drop it off. I'll get to it in maybe a year." He didn't have that kind of time. Eleven months later, the 935 EV was ready to go.

Bisi rolled the car out behind the shop for its first drive. "I got to half throttle and almost urinated on myself. I was screaming in the car. I did it again and again. I ran inside and grabbed my team and made each one of them drive it.

"It was what I have been missing," he said. "So overnight I became an advocate, much to the chagrin of my audience. I lost several supporters of my turbo work, but the [EV] feeling was unlike anything I have ever experienced."

The car was finished in a shocking pink and metallic gray combination created by Andy Blackmore. Bisi loved it because it looks like an update of a period correct 1980 935 K3 Gozzy Kremer.

The car was launched at SEMA in 2019, and needless to say, it was an incredible hit.

Shortly after SEMA, Bisi was contacted by German fashion designer Errolson Hugh of Acronym clothing fame and car designer Kyzyl Saleem. Together they came up with a show car concept called Moby X.

After seeing the initial drawings, Bisi agreed to create it.

It is now the very first 935 Long-tail (comparable to the 935/78 *Moby Dick*) road car ever approved by the California DMV as well as being 50-State legal. It's the perfect embodiment of Bisi: always curious, always improving.

TOP LEFT: This dead-on rear shot of the Moby X reveals the full effect of the Kramer body grafted to the 911 body lurking beneath. Just like the original race car.

TOP RIGHT: Bisi Ezerioha in his shop where he hosts a social-media live feed every Tuesday sharing his experience and knowledge. His way of changing the status quo is by being open and honest with the enthusiast community.

BELOW: Singer's attention to detail is evident all the way from the front bumper to the back and everywhere in between.

BOTTOM: Could the fact that Singer Vehicle Design's founder's roots are steeped in the R Gruppe give his brand more street cred? Perhaps. For example, here's Dickie Meaden driving "The New York Commission" on California's Coast Highway.

SINGER VEHICLE DESIGN

"A Relentless Pursuit of Excellence" is Singer's latest tagline and quite on target. Since 2009, Singer has been "reimagining" 1989–1994 Porsche 964 Coupes. The company's global success has led to a new definition of their brand, now known internationally as a "California Luxury Brand." Kudos to California for once again being ahead of the automotive curve as well as being the continuing home of hot-rodding—even with Porsches.

The brand is the brainchild of former Catherine Wheel lead singer/guitarist Rob Dickinson. The UK-based band's music is categorized as alternative rock, and they released five albums over 1990–2000. Music videos made back in the day all feature camera angles accentuating his intense eyes. Perhaps this foreshadowed the intensity with which he pursued Porsche perfection.

The Singer motto has always been "everything is important." No detail is too small. From exterior paint, to the tiniest interior detail, to the 911's frunk, this focus is evident throughout every build. Each element is jewel-like. The attention to detail on every level is inspiring and thought-provoking.

Each Singer is unique and designed as a collaboration project with each owner. The personalization process is extensive and specific to Singer. It's the opportunity to construct the 911 of your dreams.

Singer currently produces three different Studies from which builds are created. The Classic Study is a normally aspirated, air-cooled car also known as the *Porsche 911 Reimagined by Singer Series.*

Singer's Dynamics and Lightweighting Study (DLS), seen here at Lüftgekuhlt 9, is based on incorporating techniques derived from Formula 1. The engine, for example, produces over 700 horsepower at more than 9,000 rpm.

Next up is their Dynamics and Lightweighting Study (DLS) with inspired technology derived through Formula 1. Singer describes this as the most advanced air-cooled 911 in the world. If you're attracted to 9300 revolutions per minute (rpm) redlines from an air-cooled 911 with twin injectors per cylinder and four-valve heads, the DLS is the car for you.

But if nothing short of a Turbo is your game, consider the Turbo Study. This unique styling study combines lines from the Porsche 964 and the 930. If you are into wide bodies, this is it. It combines the best of both classics to create an exclusive Singer look.

Add a Mezger 3.8-liter (231.9 cu in) twin-turbo motor producing north of 450 horsepower (331 kW), a six-speed transmission, and your choice of rear-wheel or all-wheel drive and you now have the ultimate 911 Turbo or GT2 setup.

Singer also creates one-off vehicles like their most recent venture with Tuthill Porsche, the Singer All-Terrain Competition Study, or ACS.

This vehicle is a radical off-road twin-turbo 911 done for a private, existing Singer client who commissioned two examples. Rob Dickinson was aware of Tuthill Porsche's success with its Below Zero Ice Driving and Rally Training, so he approached them with the project. The resulting cars are brilliant. According to Richard Tuthill, working with Singer was a wonderful collaborative experience. Singer's attention to detail and design combined with Tuthill's rally experience was quite the winning combination.

Performance is paramount to any Singer. The name has become synonymous with perfection in both design and build quality as well as performance.

The result is even more impressive when you consider Porsche's own long history with rally racing. In fact, the first race entered with a factory-prepped 911 was the 1965 Monte Carlo Rally.

The two rally cars are an insane combination of design and engineering. Complete with full carbon fiber bodies redesigned to feature full clamshells both front and rear, they are capable of completing anything from the SCORE Baja 1000 to the Dakar Rally.

Singer cars are, arguably, the ultimate Porsche Outlaws produced today with no end to their creative endeavors in sight. That's not bad for someone who cut his teeth as a former R Gruppe member.

Built from a donor Porsche 964, this early *New York* car (named after the State where it was commissioned from) is backdated to appear more like an early 911. They refer to this as *reimagining* the car.

Accidents happen, but sometimes they create something you never would have created on your own. In this case, it's an undeniable signature on the rear of Rob Ida's 356.

ROB IDA

Rob Ida has built some of the most iconic and celebrated hot rods of all-time. His work as a master coachbuilder is known around the world. So why is his 356 Coupe his favorite car?

"I've put thousands of hours into restyling and redesigning cars, and yet that car is the car that decorates my home the most," Rob said. He literally has everything from photographs to paintings to sculptures throughout his home all dedicated to the car he calls *The Eight Ball*.

Rob has built everything from period correct 1940s Willys Gassers to a SEMA Best in Show winning, completely restyled 1940 Mercury Coupe. His creativity is boundless, and his execution is miles ahead of the competition.

"That's the car [his 356] that's the most fun to drive," he explained. Rob is game for anything in that car from driving to work on a Wednesday morning to doing some sort of motorsports event with it. He's even driven the 356 to The Race of Gentlemen (T.R.O.G) where he was showing a national-winning show car, and his 356 ended up drawing as big a crowd as the actual show car. "The car just checks all the boxes. It has just done everything without even trying. That's what's so charming about the car—not to mention it has never let me down."

TOP: From the front, this could be any other red 356 with an eight ball on the hood. There are so many. . . right.

LEFT: At speed, it's easy to notice the rake of the stance. Nose down like a true hot rod, *The Eight Ball* looks like it's in attack mode constantly.

ABOVE: One way to disrupt the community as a hot-rodder is to show up at SEMA with a modified Porsche 930 Turbo, especially when you won the previous year with a '40 Mercury.

OPPOSITE, TOP: In an effort to get the nose lower on the 930, Rob Ida did the unexpected by completely redesigning the car's front valance. Some thought it sacrilege, but most found it brilliant.

OPPOSITE, BOTTOM: Dead-on profile shows the 930's true stance. Notice the nose down attitude like *The Eight Ball* and the way the tires are tucked into the wheel wells. This visual detail takes countless hours to get right.

One of the more recognizable features of the car is a battle scar earned in Ida's own driveway. One morning while shuffling cars in the drive, Ida walked away from the 356 without engaging the emergency brake that needed one additional click. It's that well-known clip we've all seen on TV, social media, or YouTube: turn your head for one second, and the next thing you know you're chasing down your own runaway car. "Fortunately," the Porsche came to a stop with the help of his wife's parked SUV.

When he returned home from his event, he banged out the damage and left it in bare metal. And so it remains to this day. That bare metal fender has become the 356's signature. It starts conversations. When it gets a little rusty, some steel wool and a thin layer of WD40 shine it right up.

In 2016, Ida decided to turn a car he was building for himself into a SEMA show car. So, rather than showing up with a hot rod or a Tucker, as he is best known for, this time it was a Porsche 930 Turbo.

To give the car *IdaTude*, he achieved the hot rod look mostly through the wheel and tire combination. If you look at the car closely, you'll notice it has smaller wheels and tires with taller sidewalls. You'll also notice the car sits lower up front, giving it a raked side profile. That's not a common look in the Porsche world. This is old-school hot-rodding through and through.

OPPOSITE, TOP: The stock Turbo whale tail combined with RS bumpers really work well together. Notice the center reflector has been replaced with an aluminum strip with *PORSCHE* etched into it.

OPPOSITE, BOTTOM LEFT: If there's one thing that is consistent in every car Ida builds, it's what I like to call *IdaTude*. Nothing rolls out of Rob's shop without a healthy dose.

OPPOSITE, BOTTOM RIGHT: Speaking of *IdaTude*, who else do you know who would race his 356 on the track one day and in the mud the next?

When it came to the body, the overall mass of the front was reduced by creating a new bumper and eliminating the stocker's accordion look. This allowed him to get the car lower up front. Basically, Ida cleaned everything up and made it smaller.

The 930 Turbo became the centerpiece of the Spies Hecker display for Axalta Paints at SEMA and was very well received. In fact, it was so well received that despite Ida having turned down an offer for the car while it was being built, someone at the show made him an offer he just couldn't refuse.

A few months after SEMA, Ida was contacted again by the same party who had initially offered to buy the car. Rob had the unfortunate task of telling him it was sold, but he agreed to build another 930 for him.

The color choice this time was a Porsche color from 1957, Aqua Marine Blue. Apparently, in 1957 the color name was used twice—once for a medium-blue metallic and then later in the year for a more rare non-metallic, darker blue. Ida had previously owned an Aqua Marine Blue 356 Speedster, and when he restored it, he retained that color because he liked it so much.

Ida's Porsche connection has gone on for decades, not only through the cars he's owned and built, but also in the materials and finishes he has chosen for non-Porsche cars he has built for customers. He has chosen both exterior colors and interior leathers from Porsche for several of his projects. Rob once built a show-stopping 1937 Chevy where he used a combination of Terracotta and Cork Porsche leather for the interior.

Ida is also a big fan of the Porsche community. After owning quite a few and restoring many, he understands that most Porsche owners are die-hard hot-rodders. They embrace modifications far more than owners of other marques. This is true even with some of the high-level restorations he has done on four-cam cars and lightweight Carrera GTS cars, where a customer might deviate from stock when it comes to exhaust or light covers. Ida agrees that super rare cars should be left alone and preserved as they were built, but production cars can be open to modifications and personalization.

Of course, there always comes a point where enthusiasts need to be careful about what it is that they are altering. Today, no one would be bothered if Ida were to cut up a 996, but one day that will change. But that is the ever-evolving nature of hot-rodding.

As of this writing, Ida is working on a 1963 Corvette split window coupe that he is totally rearranging in terms of proportions and other points. He has received more enthusiast backlash on that project than he has ever received from any car community.

"Everybody knows I'm a big Willys fan, and finding an unmolested '40 Coupe is now the holy grail because we cut them all up and made horrible race cars out of them."

I asked Ida if he thought Ferry was a hot-rodder.

"I do believe Ferdinand was a hot-rodder," Ida agreed. "His Dad just helped him build the thing. That first car was really a hot-rodded Volkswagen. That's what a Porsche is; it's a hot rod! So, that's why I never feel bad about modifying them."

BRUNCH

The depth of the Aubergine paint creates a complexity similar to that of a red burgundy wine in a Riedel crystal glass. It's so rich you want to touch it . . . but know that you shouldn't.

THE DRIVERS

I handed Patrick my business card which features an image of a '32 Ford. He was dressed in a DHL racing suit and piloting the Porsche RS Spyder for Penske Racing while telling me about his '34 Ford project. He's a hot-rodder through and through.

PATRICK LONG, COMIN' IN HOT!

If you're reading this book and you don't know who Patrick Long is, you need to put this down now.

Okay, I was kidding. Here's a condensed Patrick Long history. Long has raced everything from go karts to NASCAR (National Association for Stock Car Auto Racing) to IMSA (International Motor Sports Association) in everything from pickup trucks to a Porsche RS Spyder. He has competed at tracks from Charlotte Motor Speedway to Le Mans and everywhere in between.

Long retired from racing at 40 after working as a Porsche factory race car driver for over 20 years. For his second act, he joined forces with Jeff Zwart and Howie Idelson to create the ultimate Porsche event named Luftgekühlt, a celebration of air-cooled Porsches. More recently, they expanded the concept to include an additional event called Air|Water, which is a more inclusive event accepting all Porsche types whether air-cooled or water-cooled.

Long is a native son of California and a true hot-rodder at heart. And yes, he owns a 1934 Ford and has for years. As a kid, he was inspired by all kinds of hot-rodders, among them the legendary Dean Jefferies. As a grade school kid, he loved to stop by Dean's place, and he has pictures of himself standing by the *A-Team* GMC van.

Long is clear in his principles even when applied to hot rods and outlaws: "I can't speak for others, but I can tell you about my directive. Where I see quality in a build is [in] innovation, creativity, all backed by engineering and performance. All show and no go does nothing for me."

Long has a few hot buttons when it comes to some of the performance cars being built these days, particularly Porsches. One is when someone has put all kinds of money into a car and throws it all away by adding modern tires and big wheels. Another is seeing a car that has been all flared out and looks badass yet nothing has been done to the car underneath. There's no performance or handling upgrades whatsoever. Another one is what he refers to as "meth head" headlights. "Modernized headlights work great . . . but they eliminate the period correctness of the car. If I ever did it, it would have to be a frosted lens." He understands that the headlights aren't any early Porsche's strongest suit. "Safety upgrades make sense. But you can do it in a really cool, engineered way."

When Long hot-rods a Porsche, he likes to do it with what he calls "race inspired correct authenticity." He explains that "Changing colors of the bumpers and putting numbers on the door, enlarging the wheels, does nothing for me. There [must be] a framework, a foundation for those decisions, and too many cars I see are just visually inspired, but there's nothing behind the curtain."

When Long develops a car either for himself or a friend, he starts from the driver's seat and works out. Starting with the touch points makes incredible sense given his hours behind the wheel. He learned long ago that you need to have a direction and a theme when building cars. If his cars have an overarching theme, it would be *sleeper*, an old-school hot rod term describing a stock-looking car that pulls alongside you at a traffic light and then blows your doors off when the light turns green. Long also likes to consider what Ferdinand Porsche, or a member of his team, might think about his build.

At the 2016 Luftgekühlt 3 event, Long unveiled the first Luftauto and brought Safari to the forefront of the conversation. "It raised a lot of eyebrows," he recalls, "but that was very early on and sort of a revolution of celebrating Porsche Safari efforts, Dakar etcetera. It's a little played out now—dare I say that?"

That first Luftauto was bare bones, with not a lot of creature features. It was all business. By the time, Long, Idelson, and a team of collaborators—E-Motion Engineering, Emory Motorsports, Benton Performance, among others—built their third car, it had a lot more comfort built in, including a radio, Bluetooth, air conditioning, and upgraded interior. Needless to say, it had a far different budget than the first car. It was a long build. "If there was never another Luftauto off-road car that would be okay with me," said Long.

The amount of time and patience getting this stance right is well worth the effort, and not just for visual satisfaction. When Patrick dials in a car, it's all about business.

ABOVE LEFT: Right down to the tread pattern of the Pirelli Cinturato 215/60 VR 15 tires, Patrick Long's car is dialed in to create an authentic purpose-built experience.

ABOVE RIGHT: One might think Long would opt for a larger motor than the 2.7-liter in his personal car. Nope. Maybe it has everything to do with the sound of the engine as the revs climb. Those mills sound downright angry.

OPPOSITE, TOP LEFT: Again, period correct Recaro Sports seats finished in black leather and black corduroy pull the interior together in a time capsule.

OPPOSITE, TOP RIGHT: Although all neat and tidy, this image cannot reflect the absolute wonderful angry noise that comes from a well sorted 2.7. There's something about a high revving smaller displacement that is just pure magic.

Long was approached by Mattel in 2020 to come up with something for their Hot Wheels Legends Tour. Having lived in Florida years ago, he thought it would be fun to build something that could withstand any environment: a Porsche that embraced the sun, the sand, and those unexpected rainstorms that come out of nowhere when you had the top down and a vehicle with crazy resilience so when it was full of pet hair and sand, you could just hose it out. He came up with the combination of a Porsche 914 and a pickup truck and a beach buggy. The car featured huge, flared fenders and a radical suspension.

The car was unveiled at SEMA 2021 in the Mobil 1 booth and went on tour for two years. There have been three different versions of it as a Hot Wheel, which were sold around the world. "It made a lot of kids happy," Long smiles.

The car was so successful that Mattel worked up a Safari-style Porsche 944, which debuted at Rennsport Reunion VII in 2023.

How about what's in Long's garage? One of his favorite personal cars is his 1972 911 T in a Porsche color called Aubergine. I will tell you that after following Long for several miles for a photo shoot—that car is the real deal. The stance, the sound, everything about the car is spot-on. Long says that's all because of the car's underpinnings. It's not about the tires, it's not about big horsepower, and it's not about aero.

Basically, the T is an M472 RS including the camshafts.

The T was always a Bay Area car with matching numbers. It was previously owned by Craig Watkins, the cofounder of Flying Lizard Motorsports. Watkins says he found it in the Sunday classifieds of the *San Francisco Chronicle*. He sold it years later to a woman named Kathy who was a PCA "track junkie." She then sold it to Porsche racer Johannes van Overbeek. Van Overbeek and his dad built it into his dream car with help from Jerry Woods before selling it to Long.

Long has since changed the seats to a RECARO custom seat. "It's a transitional seat between a classic 70 F body Sport Seat and what evolved into the early G Body square top sports seat. So, it's an original development seat which would have been what I believe to be around '73."

I asked Long what he thought about the Dean Jeffries Outlaw.

"I love that cross pollination between a hot rod/studio movie car builder and a Porsche. I love it when someone from outside the normal realm comes in and does a one-off or a small series," Long answered. "I seem to have more of an appetite for the art—or what some would call questionable decisions. Like when Rob Ida built that 930—because that's really pure art." He went on to comment how Rob Ida "nailed the tire stance, the tuck of the rear, and how [the body] sits over the tire that is so frequently missed."

There are of course certain things that are off-limits when it comes to outlawing in Long's view—limited-run production cars for one. He believes the condition of the car should play into any build decision. If it's a really well-preserved survivor, it should be left alone. "I hate it when someone chops up a really clean, well-preserved car."

He commends his friend Rod Emory for breathing life back into Porsche carcasses that had been left for dead by reimagining the cars and restoring all the metal.

Long thinks if you are altering a car in any way, like backdating or pushing it forward, you should be able to reverse the work down the road. All the original parts should be retained. On the other hand, as he notes, "If they're out there being driven, who am I to be the Fun Police?"

Tucked neatly inside the fender wells are these subtle louvered liners lending that understated hot rod look.

Magnus Walker, a.k.a. "the Urban Outlaw," and his dog Willow chilling in the Arts District of downtown Los Angeles among his vast Porsche collection.

MAGNUS WALKER "THE URBAN OUTLAW"

I had an opportunity to visit the "Urban Outlaw," Magnus Walker, and ask him how he defined the moniker *outlaw.* "Outlaw is a term I didn't give myself. It was from an article in *Total 911 Magazine* from 2011 where they termed me as an 'outlaw living in an urban environment.' To be honest an outlaw is just a state of mind. It all started with James Dean and the 'Little Bastard.'"

Walker has been into Porsches since he was ten years old. He bought his first Porsche when he was 25, and he has owned and modified them for the past three decades. "The very first Porsche I bought was modified," he said, recalling his 1974 Porsche slant-nose conversion.

He bought his second Porsche, a 1971 911 T, in 1999 at the Pomona Swap Meet for $7,500. It's the car now known as the 277 Porsche. Walker built that car into a 1973 RS replica. Everything on the car was state of the art about 20 years ago when he first built it, and it's been modified several times over the intervening years. He never had an engine built specifically for the car. Instead, he would source race motors from others who were upgrading. Walker's 277 has been variously powered by a 2.4, 2.5, 2.7, and a 2.8 twin-plug engine.

"The great thing about the 277 is it's a 10/10ths car. [At] 9/10ths on the street—foot planted all the time—it doesn't really have so much power that it's unusable. It is a usable performance. You know it's not the quickest car I own but it's the most engaging car. It just checks all the boxes."

The car became a celebrity after the launch of Walker's 277 film *Urban Outlaw* in 2012. Suddenly, there were countless magazine articles, and it seemed everyone wanted to drive it. In 2016, Hot Wheels created a model of the car, and there are now eight versions of the model sold worldwide. Walker's celebrity status upgraded when 277 became part of the video game *Need for Speed* and the *CSR Racing 2* game. Nike even did a collaborative sneaker based around the car.

Today, the 277 is not being driven as much as in days past. It's too much of a celebrity and draws too much attention. "Life's interesting when you start driving your 2014 Turbo S as a commuter car, [so] you can blend in, ironically, and fly under the radar. I don't always want to attract that much attention and heat."

Walker thinks it's flattering when someone builds a replica of the 277. He even did the "Tokyo Outlaw" video in a replica everyone mistakenly thought was his car.

Of course, the 277 is not the only Porsche he has built. Walker's favorite car is his 964 with a channeled hood and louvered fenders. He remembers a time when no one cared about the 964, and you could buy an RS America for thirty grand. Though the 964 is one of his favorite builds, it does not have the notoriety of others because he has never featured it in a film.

Walker isn't a participant in the Porsche club scene. He is proud of the fact that all of the events he has created, from here to Tokyo to Sydney to London, are open to all. They are not Porsche specific or cliquey. Walker's relatability extends beyond the Porsche world because like other enthusiasts, he's a builder, a collector, and a driver.

"Porsche is the common bond that brings like-minded people together from all walks of life," explained Walker. "You can be hanging on the lawn at a concours talking to a billionaire that you would never normally hang out with, but you've

Perhaps the most famoust Outlaw of all time the Magnus Walker 277. You can even buy a Hot Wheel of it!

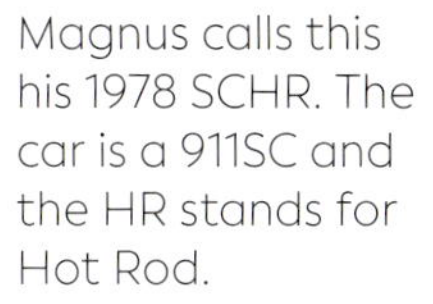

Magnus calls this his 1978 SCHR. The car is a 911SC and the HR stands for Hot Rod.

Magnus's passion runs deep for all things Porsche. His personal collection continues to expand, and he is always searching for the next significant addition.

got a common bond talking about the love affair of this thing called Porsche. It's the great equalizer."

Walker is less interested in someone who has a Porsche with just 100 miles (161 km) on it and has much more in common with someone who has rolled 500,000 miles (804,672 km) on their car.

He grew up with a lot of American TV shows like *Starsky & Hutch*, *The Dukes of Hazzard*, *The Rockford Files*, and *Kojak*. Stir in a little sex, drugs, and rock 'n' roll, and you'll start to understand his background.

Walker and his late wife, Karen Caid, were known for their clothing line Serious, based in the L.A. Arts District and that catered to the rock 'n' roll and celebrity scenes. The brand received considerable editorial coverage as people were actually *wearing* their shirts while on the cover of *Rolling Stone* and *Spin*. They were outfitting everyone from Alice Cooper to Madonna and selling wholesale to 500 retail stores as well as having their own store on Melrose Avenue in Los Angeles. The Walkers built their business pre-Internet, but managed to receive editorial coverage in outlets like *Vogue* and the *Los Angeles Times*. The exposure and limelight not only prepared him for what was next, it also made him comfortable in front of the camera.

Serious' success helped Walker collect some of the cars he recalled from the American TV shows he'd loved as a youth. His early collection included a Plymouth Super Bee, a 1965 Ford Mustang, a 1967 Jaguar E-Type, and even a '73 Lotus Europa, all populating his garage well before his obsession with Porsche and the Urban Outlaw brand.

Walker has lived in the United States for 37 years now, and everything he has pursued and accomplished was done without specific instruction or experiential background. It's all down to tenacity—figuring out how to design clothing, how to restore a building, and how to build some of the cars he owns. The common thread here is that Walker has been true to himself and passionate about everything he's done. "It's all sort of based around expression, style, and aesthetic," he explained.

He bought his first building in the L.A. Arts District 24 years ago, and everyone said he was crazy. "It turned out to be the best thing we ever did." He trusts his gut when it comes to any decisions. He relies on no one else's approval or opinion. Walker has seen many of his friends derailed by other people's bad advice. "If you don't need validation from other people, and you believe in yourself, fuck it, just go do whatever you want. You'll find a way to make it happen."

Walker has once again been expanding his collection to include many other marques alongside his beloved Porsches. His passion is broader than just one manufacturer. It's also exemplary of the relatability of his personal brand. Walker is all about hot-rodding, and that tendency is global and connects all enthusiasts.

Walker has suggested we may see a new collaboration in the future, something in line with his broadening collection. He thinks our response will most likely be, "Well, I didn't see *that* coming."

ABOVE: There's always a fresh project underway in the back room with plenty of parts available for inspiration.

RIGHT: With a collection as large as Magnus's, there's always something that needs attention. In this case it's new brakes for hit 1968 911 RST.

OPPOSITE: Magnus's rock 'n' roll roots are always just below the surface and readily available at a moment's notice. From his look to his attitude to the cars he drives . . . "It's all rock 'n' roll, baby."

OUTLAW
PORSCHE

Keens can often be found lurking near the docks of his childhood home of San Pedro, CA. As a native, he'll make sure you know how to pronounce it correctly as well.

THE DAVID KEENS HOT ROD

Sometimes to understand an outlaw car, you need to understand its inspiration. David Keens and his Porsche 911 are a perfect example. Many of David's earliest memories are based around his father and his Austin-Healey 100-4. In fact, in one of the earliest photographs of Keens and his father they are standing right next to that car. "Ever since [that] photograph from 1955, it's just been my dad teaching me about cars—all kinds of different cars."

Keens recalls sitting with his father listening to the Indianapolis 500 in 1959 when it was only broadcast on the radio. He and his brother would sit in silence as his father concentrated with pencil and newspaper in hand crossing out each car that dropped out.

Growing up there were only three magazines in the house: *Road & Track*, *MotorTrend*, and *Hot Rod*.

Keens and his father attended hot rod shows in Los Angeles and sports car and Canadian – American Challenge Cup (Can-Am) races at Riverside. They would drive to the races in that Austin Healey 100-4, park it in the dirt, throw on the tonneau cover, and then spend hours watching the races through the esses. It was there that Keens' father helped him differentiate audibly between a Ferrari, a Maserati, and an American V-8.

"If you want real louvers . . . you go to a hot rod shop." David describes the look and feel of his rear deck lid louvers as a lost art.

The big noise didn't stop there. Keens' father became a member of the Lions Club in San Pedro, California, the very chapter responsible for creating Lions Drag Strip in an effort to cut down on street racing and create a safer place for teens to race. Keens remembers growing up in and around the drag strip, his weekends spent surrounded by Willys Gassers, Ford Anglias, and rail dragsters.

When you look at Keens' Porsche, you start to see bits and pieces of his automotive history in its build. From the rear-deck louvers to the tire sidewalls, even the paint colors, it's all reminiscent of the cars he saw as a child.

Keens has owned several Porches throughout his life and has had the opportunity to race them as well. In his racing days, he worked with Jim Buckley and his Fort Worth, Texas, based shop Buckley Racing LLC. It was at Buckley that he learned "the car has to have a little rake to it." Buckley thought rake helped a 911 to handle better on the track, and that became his signature setup. That's why Keens' car sits the way it does. It's also reminiscent of the cars he experienced at Lions Drag Strip.

When Keens decided in spring 2017 to build a hot rod Porsche, he wanted to base it on a 1972 911 S. But by that point, early-car values had accelerated beyond his budget. His donor car became a 1979 911 SC bought from Paul Kramer at AutoKennel.

Kramer's father, Ed, had lived across the street from the SC for years. The car was seldom driven. It was a one owner, perfectly preserved car in Grand Prix White. Keens told Kramer, "You know I'm gonna cut this thing up?" Kramer's father was beside himself every time he saw Keens' in-process documentation photos. In the end, however, Ed was pleased with the final result.

What Kramer and his father didn't know is that Keens is a true artist. Literally. Art has always been part of his interest in cars. "I can't separate them [art and cars]. It's all the same—going to custom car shows, going to hot rod shows, going to races—there's a visual aesthetic to each of those cars, and they all have a certain aesthetic based on their purpose."

Growing up, young Keens was interested only in cars and art. He wanted to be an artist from the time he was a child. He feels fortunate that his parents supported his artistic interests every way they could. They took him to museums and art installations to study. Keens studied at California State University, Long Beach and did his graduate work at University of Washington, Seattle. From there, he became a successful artist and eventually moved on to a career in academia. He taught at the University of Texas at Arlington in Dallas–Fort Worth where he stayed for 39 years. His 911 may be the only Porsche Outlaw built by a professor emeritus.

Across Keens' career, his mediums evolved from large sculpture to small sculptures and eventually to glass. He describes the small-sculpture phase as an intense appreciation for working with his hands and attending to very small details, somewhat like working on a car or building a model. In his art's next phase,

ABOVE: At speed, David's car hunkers down and gets the job done with the help of his Group 4 Compagnolo wheels all tucked neatly within the fenders. Not to mention the help from his StopTech oversized brakes.

OPPOSITE, TOP LEFT: 1967 911 R taillamps are popular with the outlaw crowd. The simple, round dual-lamp design with frenched mounting screams hot rod.

OPPOSITE, TOP RIGHT: Precision cut louvers all lined up perfectly, fitting in with the form follows function ethos of Porsche.

OPPOSITE, BOTTOM LEFT: Yellow headlamps, or "yolks", have become popular in the outlaw movement. They can look out of place at times depending on a car's color scheme, but fit right in with this custom paint color that the owners call "Fresh Custard."

OPPOSITE, BOTTOM RIGHT: A pure analog interior defines and focuses the driver's intention. Keen's office is all business.

Keens moved from tedious, intimate, intricate metal work to fluid, spontaneous, direct glass work.

Keens has been driving Porches since 1975, and his hot rod is a reflection of his past. It's a culmination of experiences colliding. From his racing experience and his creativity as an artist, he knows that all the details matter. Custom cars, race cars, and traditional hot rods each flow from a certain concept, and each of their respective details should enhance or complete that concept.

With his Porsche Outlaw, Keens wanted to create a true hot rod 911. He wanted it to be modified for a particular look and with a particular performance aspect. The look he was after was "kind of the slightly pissed off hot rod. Not over the top, just kind of a bad boy on the street that you really [wouldn't] want to go over and talk trash to because he'd probably kick your ass. But he's not going to come over and kick your ass if he's not provoked."

"I've spent enough time on the track with 911s and had enough different models to know what I don't like," he said. "I just did what I wanted to do. Better performance and aesthetics. It's just what I wanted. It's what I think the perfect 911 should be."

Aesthetics and design is what it's all about. Now, it's clear why it took months for Keens and his wife Arianne to choose the right color for the car's exterior. They spent hours looking at different whites, trying to come up with the right shade of what they call "Fresh Custard." They began to understand that many commercial whites have differing amounts of gray in them. They needed to tone down the gray and add just the right amount of color to get what they were looking for.

Once the white was right, next came the metallic gold of the front deck lid. If you look closely, you'll see a heavy amount of black with green added and a very fine metal flake.

Keens' car reveals subtle details everywhere you look. From the finish on the wheels to the headlamp lenses, you need to look closely to spot all the subtle changes. One of Keens' favorites mods is to the wheel arches where the flat edge has been rounded off and removed. It's so perfect it looks like it's OEM.

"Design wise I just couldn't get my head wrapped around that flat edge," explained Keens. "Why would you do that? By eliminating that and giving it just the right amount of curve, to me . . . It's a big deal."

He was also adamant that the rear deck lid needed to change. Every good hot rod has louvers. Come to think of it, so did that Austin-Healey. So, it was only right that Keens take the car to the Shine Speedshop and ask Jimmy Shine to louver the deck lid. He remembers when he described to Shine what he wanted to do to the Porsche 911, "Jimmy's eyes lit up, and he agreed to put the louvers in himself."

Keens loves how the deck lid turned out. "He did a fantastic job. It was terrific, so I just decided to leave it in bare metal." People ask all the time if the deck lid is clear coated. Keens laughs that "I don't do anything." In fact, that bare metal deck has become a signature of the car. There are no plans to paint or clear coat it—ever.

Sooner or later folks will ask you to lift your Porsche's deck lid especially when you have an aluminum one with custom louvers. One day when the car was at Revival Road Co. in Monterey, California, awaiting paint, it was suggested to Keens that the original 120,000-mile (193,121 km) 3.0-liter (183.1 cu in) motor might

not be something you'd want to show anyone when asked to lift the deck lid. For comparison, a technician at Revival lifted the deck lid on a few other cars in the shop to show him what it could look like. After Keens' wife saw the options, she agreed it was time to address the engine. The end result is a 3.4-liter (207.5 cu in) short stroke with electronic fuel injection (EFI) and individual throttle bodies, AEM Infinity control system, twin-plug heads atop 10.5:1 compression pistons, all internals in titanium—all the good stuff. The intake is essentially a replica Porsche RSR high butterfly. Everything is clean and hidden to give it the full RSR look.

Open the doors, and Keens' interior is all business. "There's nothing fancy about it," he says. For him, the most important thing is his seat position. "I love doing 400 miles [644 km] on a day trip. I like to go up past Ojai and beyond and back. Those are the days I really like, so it's gotta be comfy." Combine GTS seats with a MOMO Magnus Walker Signature Series steering wheel and it's a perfect combination. Keens loves that steering wheel in particular, enthusing that "it has just the right amount of dish, perfect diameter, and just the right amount of patina."

If you're ever passed in the canyons of Angeles Crest Highway by a screaming 911 with a bare-metal deck lid, you'll know it's David Keens exercising his angry hot rod!

Armed with ample torque and horsepower, this Porsche SC Outlaw feels at home on streets or track.

Industrial design is Magnusson's life's work. No wonder his 912 Outlaw stands out against all the rest. It's the perfect combination of badass and style.

CARL MAGNUSSON

One of the great things about the car world is the people you meet. The automobile is the conduit that brings us together. You just never know who you're standing next to.

Or in my case, you never know who you're parking next to.

In the early 1990s, I parked my 911 SC in an outdoor lot in New York City's (NYC's) Soho district, right on the corner of Mercer and Prince across from Fanelli's. I noticed a 356 cabriolet parked there occasionally. It had some unique louvers on the rear deck lid and a rather interesting rear valance. The car was clearly customized and the attention to detail was evident throughout. One day, I placed my business card on the windshield and asked the parking attendant to please have the car's owner call me. A few days later, I received a phone call from a gentleman named Carl Magnusson. Our conversation was brief and wrapped something like, "Certainly you can photograph my car. Your [business] card is gorgeous and by the way, who the hell are you?"

Magnusson and I set up an early morning shoot on a Saturday. I asked him to meet me in the back alley behind the Puck Building on Jersey Street. I chose this location based on the early light and the Puck's architecture, which featured large, shuttered windows that seemed so out of place in NYC and lent it a European feel.

Fast forward decades later, and I stumble across a post on Instagram and an exchange between Magnusson and Rod Emory. I reached out to Rod and verified this was the same Carl that I had met many years ago. Apparently, Emory had built Magnusson's 356 cabriolet that I had photographed years before. Talk about six degrees of separation! I reconnected with Magnusson and learned he had just completed building a new Porsche Outlaw based on a 1968 912.

Let's step back a little and talk about who Carl Magnusson actually is. Born in Malmö, Sweden, he and his family moved to Canada when he was a boy. Imagine him on a farm, a place he described as "a pretty desolate part of the world in between Calgary and Edmonton" where he was surrounded by farm equipment that he recalled as "amazing machinery."

At the age of 15, Magnusson subscribed to *Rod & Custom* magazine. He was fascinated by the modifications he saw on its pages. From frenched headlights to smoothed hoods, chopped and channeled cars, "I absolutely loved it all."

Because Magnusson lived on a farm, he was eligible for his driver's license despite his tender age. He remembers going down to the Alberta Department of Motor Vehicles, which doubled as a general store/liquor store in his small town, to procure his ticket to motoring.

"One day I was given $50 to go do something with the understanding that maybe there might be a little bit left over for me," he recalled. Errand bound, he was distracted when a friend said, "Hey, there's a 1934 stove-bolt Chevy Coupe for sale. We should take a look at it."

Every angle of this car reveals subtle touches only noticed by true aficionados. Just like any great hot rod, one must look closely to see the alterations from stock. Note the absence of side view mirrors, larger wheels and tires, and the added louvers in the rear valance on either side of the license plate.

ABOVE: Great design just works. Straight lines and curves intermingle like contrasting patterns, creating a visual resting place for the viewer.

RIGHT: From the most elegant architecture of Manhattan to the street art of Brooklyn, the car looks at home no matter the backdrop. Take a moment here to squint and see if you see the faces.

Just like that, the $50 was gone, and Magnusson was soon chopping and channeling his own car in a machine shop. "If you look closely at [a] photograph my sister took, you'll see I could cut, but was not good at welding."

Magnusson's inner hot-rodder couldn't resist the opportunity to drop in a larger motor. He remembers, "I couldn't afford the driveshaft, so I shoved the motor and gearbox back into the frame until it hit the diff. I connected the gearbox directly to the differential via a u-joint!"

While ripping around the farm on dirt roads he thought, "I'm gonna need a helmet," so he ordered one through *Hot Rod* magazine. After receiving it, he repacked it and shipped it to Von Dutch with a $5 bill and a note asking Dutch to pinstripe it. Amazingly, he got it back with hand pinstriping. Unfortunately, the helmet was lost to the passage of time.

For years, Magnusson would write requesting brochures from various automotive magazines. They became inspiration for hours and hours of his own automotive renderings. Eventually, he built up sufficient courage to send a few of his drawings to a magazine for possible publication. "Thank you very much for your cartoons" was the reply. Thus ended his dreams of becoming an automotive designer.

Magnusson left the farm for the University of Idaho followed by The Chalmers Institute of Technology in Gothenburg, Sweden. In 1965, he joined the office of Charles and Ray Eames in Los Angeles. From there, he opened his own design studio in North Hollywood in 1970, and in 1978, he went to work for Knoll as Director of Design where he worked until retiring in 2005.

After selling his 356 cabriolet through a broker in Switzerland, Magnusson owned several other Porsches, including a 356 Speedster. He describes the latter as "rather barren and not as much fun as [the] cabriolet. The best thing about the cab was the family could join you in the car—kids in the back, top down, great fun."

Magnusson decided to get a coupe. Plenty of his friends had 911s, and yes, they were fast, but a 912 is 1/3 of the price. Plus, he'd built vast knowledge of the Volkswagen/356 motor and felt it was plenty of power. Magnusson knew he could get just as many tickets with a smaller engine. Essentially, the 912 is powered by the last 356 SC engine, making it very reliable because it has all of the flat-four's final updates.

Magnusson found the car in New Hampshire, where he describes the previous owner as "presumably honest, [but who] apparently didn't know anything about rust." However, the car had an incredible motor built by Duane Spencer in northern California, which included upgraded 1720cc Shasta pistons, Carrillo rods, and Weber carbs!

Magnusson's 912 presents all kinds of details. It's obvious from all angles that every detail was considered in the build of this car. From the headlamps to the rear deck lid right down to the grille over the shift linkage, nothing was left untouched. However, in true hot-rodder fashion, everything looks OEM.

Removing the grille and adding louvers wasn't an original idea, but as Magnusson points out, "With the 912 motor it has no effect whatsoever on the cooling."

Magnusson also resisted changing the 912's stance or track. The car rides at factory-spec height.

The car now resides in Europe and is driven regularly to events and entered in many competitive rallies, like the Bernini Gran Turismo, as time allows.

Does Magnusson have any intention of building another Porsche?

"I think this will get me through the next decade," he says with a smile.

CLOCKWISE FROM UPPER LEFT: (1) Maintaining true Porsche "form follows function," the subtle placement of modified louvers in the rear valance help reduce heat in the engine bay. (2) The simplicity of the 356 motor in a 912 looks more elegant by design. The power-to-weight ratio is more balanced as well when compared to a period 911. (3) An icon of industrial design, Carl Magnusson, is seated in one of Porsche's legendary designs. (4) Yolks give the 912 a more European look and feel. Sighting increased visibility in glare-like conditions and reduced eye fatigue, yellow headlamps were deemed mandatory on all motorcars in France in 1937 (but are illegal in the US, so don't get any ideas).

(continued on page 126)

COLORING INSIDE THE LINES

Even though Porsche Outlaws have existed for decades, we're all aware that not *every* Porsche enthusiast embraces this approach. The thought of altering a Porsche in any way sends chills up and down their spine. Using anything other than OEM parts with a Porsche is considered sacrilege and should never be tolerated.

FOR YEARS, PEOPLE HAVE over-restored classic Porsches to a degree beyond the original spec from paint to fit and finish. You'll find examples at any concours you attend. Such cars are spotted easily by those in the know and are wasted efforts for viewers who don't recognize the difference. Perhaps the saddest thing is a Porsche that now spends its life being driven to and from events in the darkness of a trailer rather than flat out through the turns as God and Ferry intended.

Because of the divergent views of original-car versus outlaw adherents, I thought it would be interesting to speak with someone who would never consider altering their Porsche in any way and offer that perspective.

PETE RITTER

Regardless of trends or what's hip at the moment, there are those who feel a Porsche in any form other than bone stock is tantamount to breaking the law. Outlaws they are not. I think of them as "naysayers."

> ***Merriam-Webster's Dictionary* defines an outlaw thusly:**
> **1: a person excluded from the benefit or protection of the law**
> **2 (a): a lawless person or fugitive from the law**
> **(b): a person, or organization, under a band or restriction**
> **I: one that is unconventional or rebellious**
> **3: an animal that is wild and unmanageable.**

None of the above describe Pete Ritter. In fact, Ritter is the antithesis of *Webster's* outlaw definition.

Ritter acquired his first Porsche in 1998, a used Guards Red 964 purchased from a dealer lot after receiving a positive pre-purchase inspection that should not have been issued. Except for his subsequent purchase of an original-condition 1996 993 Coupe, which he still owns today more than 28 years later, Pete has only owned new cars thereafter.

Since then, he's owned a couple of 997 GT3 RSs, and currently owns a 987 Boxster Spyder, a Cayenne Diesel, a 991.1 GT3 RS, and a 991 Speedster with the Heritage Design Package.

"By the book" might describe how Pete manages his current Porsche stable. "Maybe it's a little of the OCD [obsessive-compulsive disorder] in me, but oil changes once a year regardless of mileage is what I do. I typically go by recommended Porsche service programs or their guidelines."

02

Tire pressure, ride height, and exhaust are all stock—each car exactly as it left the factory. Even the orange side marker lamp lenses that most everyone tosses in a drawer are never replaced with the more desirable "European" clear lenses.

Ritter holds onto his Porsches a long time, so he pays particular attention to tire date codes. "It's so dangerous to have old tires on your car. Safety is the one place never to compromise." By the book.

I asked Ritter about his thoughts on aftermarket products and modification, and he offered a very simple answer: "In my view, Porsche typically gets it right."

He pointed out an example of a friend who decided his Cayman R's brakes weren't powerful enough. Aftermarket brakes were installed, and this one change caused all kinds of issues with bearings, tire wear, suspension, and strain on other components. "In the end, the headache that followed just wasn't worth it at all."

Ritter is fully aware of all the countless ways one might change the performance and appearance of a Porsche. He's also aware of something that many owners neglect to consider: the car's warranty and the ramifications alterations may have on it. Therefore, Ritter is generally not comfortable altering a car from its original spec. "Quite simply," he notes, "I never want to be anyone's guinea pig."

Ritter cringes when a beautiful and all-original "pure stock" car is "butchered" to create an outlaw. He concedes his view is on the extreme side and that he's in a minority when it comes to this perspective. He can live with modifications that are reversible, such as wheel and tire changes, and explains that "I can appreciate it when Rod Emory takes what would otherwise be junk or scrapped and saves it by converting it into something that is new/fun."

All that said, Ritter does think that the Bruce Jennings 1958 356 Speedster is wickedly cool. "Jennings nailed it. It even inspired the Porsche factory Heritage Design pack for the 991 Speedster many years later. It's all a matter of taste but sometimes the owner just misses the mark."

While Ritter can't quite put his finger on it, he says, "Some outlaws are cool, and others just hurt me."

Porsche's primary appeal to Ritter is that the cars are already engineered to the hilt in a "form follows function" vein. He believes he would never have enough time to do his own research to improve on the factory's work. "Porsche is already developing things properly, taking into account all of the compromises necessary to make a car roadworthy such as safety, overall performance, and longevity. How can I improve on that?" One could argue that simple cosmetic modifications can be fine and a means to express oneself. But it is a slippery slope. Ritter simply enjoys driving his cars as they exist, including on the track, and the simplicity of washing and cleaning them, which he finds relaxing.

Of course, the degree to which one embraces car modification can also come down to personal time restraints. Pete has a demanding job, a wife and family including three children in college, and two dogs. Do the math on all that and take into consideration that Ritter doesn't really care to tinker with his cars, and you start to understand his views on radically modified Porsches.

ABOVE: Some details are better left alone, like this Sports Chrono Package dash-mounted clock and stopwatch combo.

OPPOSITE, TOP: Complete with one of a kind graphics specific to the car, the 911 Heritage Speedster stands out everywhere it goes.

OPPOSITE, BOTTOM: This 911 Heritage Speedster is a perfect example of a car you don't want to alter in any way. It's an extremely rare, limited-edition car. Every angle of this car is unique.

Anyone who can slip the phrase "no can do" into a hit single is brilliant! Sorry, I just had to state that.

(continued from page 122)

JOHN OATES

Many people know John Oates for his music, as half of the musical group Hall & Oates, but don't really know how much of a car guy he is. Oates' connection to Porsche history is extensive. In fact, in the 1980s, John raced a Porsche 924 GTR competitively. His talent behind the wheel landed him a factory driver seat for Pontiac in the IMSA GTU class driving a Fiero with veteran codriver Bob Earl.

An unfortunate racing incident at Wisconsin's Road America forced Oates to make a decision between racing and music. He chose the latter and left racing in his rear-view mirror. But you would never know it when John is behind the wheel—the driver in him is very much alive and well. It's part of his DNA, as is Porsche.

Oates' first car lot visit was with his father at the age of four to a Volkswagen/ Porsche dealership. He saw his first sports car, a Porsche 356, which may have imprinted on him. His Porsche dreams became a reality in 1976. While on tour in Hollywood, he spotted a Guards Red 930 Turbo in the window of the Porsche dealership at the corner of Wilshire and La Cienega. Oates went in to buy the car but was told by the salesman that Rod Stewart was interested in it as well. He later told his manager about the car, and the two of them returned to the dealership. Sonny Bono's 356 Cabriolet was also on the lot, and his manager was interested in that car. Oates' manager and the salesman disappeared for a while. When they came back, it was apparent that the salesman could not turn down a two-car deal—Rod Stewart notwithstanding—and that Red 930 Turbo was now Oates'.

Once Oates completed his tour in Los Angeles, he drove the 930 cross-country to New York City where he was living. The following year, he drove it back to Los Angeles for another project. He describes those rides as "two of the most epic drives I've ever done in my life."

Through the years, John has owned several different Porsches, including a 356 Speedster back in the 1980s. He still regrets selling that car.

In 1983, Oates placed an order at Porsche in Zuffenhausen after completing a private tour of the factory. That car was a 1984 G Body 3.2 Carrera with numerous Special Wishes items including a Pearl White Metallic paint job, custom wheel center caps, thicker steering wheel, sports seats, and other exclusive items. John sold that car in 1991.

In 2020, Oates' wife, Aimee, spotted the '84 Carrera advertised in an online auction as "John Oates' former car." Sure enough, it was his car, and it's now back in his driveway.

So, why does a literal rock star with racing experience who could buy any car he would like decide to build an Emory Outlaw 356 convertible?

As car guys do, Oates found himself surfing the web one day where he spotted a photo of one of the Rod Emory Outlaws posted on Pinterest. He made note of it, and the next time he was in Los Angeles, he gave Emory a call. It's as simple as that—well, sort of.

Recall the aforementioned 356 Speedster as "the one that got away." An Emory Outlaw presented an opportunity to both get that car back and create the version of his dreams at the same time.

While still in Los Angeles and after their call, Oates and his drummer stopped by Emory's shop. This provided an opportunity to sit down, face-to-face, and hear Rod's story. "The first thing that struck me was that he had this Southern California hot rod DNA [along] with this incredibly sophisticated and comprehensive knowledge of Porsche history," recalled Oates. "That combination really appealed to me."

After a spirited test drive in an Emory Outlaw, Oates turned to him and said, "I'd like to collaborate on a project with you."

For the next 2½ years, Oates and Emory were in a tight collaboration. Oates visited the shop at least once a month, if not more frequently. Sometimes, he was in town on business, and sometimes, it was a trip specifically to see Emory to discuss the project.

I was curious about Oates' choice of exterior color. He told me that it all started with the interior. He and Emory started from the inside and worked their way out. First came the leather choice, then the carpeting, and so on.

Oates also explained how they decided to create a "greatest hits" 356. Walk around the car and inspect it closely. If you're good, you'll see different aspects of the best elements of a 356 A, a 356 B, and a 356 C, but all in one car.

The smoky brown paint with black accents and wheels makes John's Oates' understated Emory Outlaw Cabriolet a real sleeper.

CLOCKWISE FROM UPPER LEFT: (1) The interior of Oates' Porsche Outlaw is nothing short of stunning. The rich brown tones of the exterior mingle elegantly with the wooden wheel and cognac leather. The combination of upscale elements combined with a race theme make it unique. Note the integrated roll bar and race belts. (2) The Emory-Rothsport Outlaw-4 Engine is Emory's own split-case engine design based on the 3.6-liter architecture. It fits perfectly in the 356's engine bay. (3) The signature reverse louvers are highly functional, pulling air out of the engine bay, and just what the car needs to cool the larger engine. (4) Oates' unique chopped, removable hardtop with plexiglass side windows stands out from the crowd. "Since we put the top on, I've never wanted to take it off," he said. "It's that cool."

Oates' racing history is evident every time he gets behind the wheel. He truly knows how to the get the best out of this special car.

The donor car was a barn find Porsche 356 B Cabriolet out of Texas, but you'd be hard-pressed to figure that out now. The cool thing was that the car came with a factory hardtop, a rare option. Emory had an idea to create something no one had ever seen before.

When it came time to decide on an exterior color, Oates suggested Graphite Metallic. Emory agreed it was a great color but disagreed on the specific hue, feeling it had too much blue in it. He thought it would provide too much contrast to the interior, so they set out to create a one of a kind custom color. They started with Graphite Grey Metallic, but instead of the bluish tones, Oates suggested they concentrate more on bronze and black. The end result is stunning. As Oates likes to point out, "Women love it; they go crazy. They all want to know what the color is because they've never seen anything like it."

Oates so enjoyed the collaborative process with Emory that I would not be surprised if the two launch another project in the future. "Perhaps the next one won't be another greatest hits car," mused Oates, "and instead be its own top-ten car."

Without making an effort to understand this 911 build, you might walk right past it. Trust me, it's well worth the time to explore in order to truly appreciate every piece of it.

THE KRONBERG PORSCHE

Before you can understand the Kronberg Porsche, you first must understand Gary Hustwit. Raised in Southern California, Hustwit was obsessed with cars, guitars, and skateboarding. He went to high school in Newport Beach and then went to college at San Diego State University. While in high school, his best friend's dad had a gold Porsche 928. To Hustwit's teenage eyes, "It was the coolest car ever." Another of his buddies drove a 914. He grew up obsessed with the Porsche 911, though he never felt like he would be in the position to own one.

At the age of 12, Hustwit became obsessed with guitars and started buying and selling them to earn money. He would comb the classifieds from neighboring towns seeking interesting six-strings. It's something that he pursues to this day despite great success in other endeavors. A true passion never leaves you. Hustwit is a walking guitar encyclopedia and can talk for hours about the different manufacturers and all the elements of crafting a guitar.

He understands how guitars age, how finishes change, and how wood needs to remain supple to last. He also appreciates that one of the most important keys to an instrument's longevity is regular play. It's the only way an instrument stays alive, through vibration. Like a car needs to be driven, a guitar must be played. It's the only way to show it's loved.

Today, Hustwit is an award-winning filmmaker who has created more than 15 documentaries. His films have been broadcast on PBS, HBO, Netflix, and The BBC in 20 different countries. Perhaps the most relevant documentary for our purposes is the one he created in 2018, *Rams*, about design legend Dieter Rams with original music by Brian Eno.

Rams' iconic designs for the brand Braun are known around the world. The ethos of his work is: "You cannot understand good design if you do not understand people. Design it made for people." At age 93, his designs continue to inspire.

Rams is also a Porsche guy and has owned 911s from the late 1960s and early 1970s as well as the 993 he purchased new in 1994 and still drives today.

After the success of Hustwit's films, including *Helvetica* and *Urbanized*, he finally found himself in a position where he was able to consider that 911 he had wanted since his skateboarding days.

But Hustwit didn't want just any 911. He had a vision. The Kronberg Porsche was not a mere afterthought, it was something he had created in his head before he even started looking for cars. Hustwit wanted to create a Porsche that looked like it was a race car from the late 1960s or early 1970s and that would have been sponsored by Braun. He wanted it to seem as though it had been found in a warehouse in Kronberg (a small town outside of Frankfurt, Germany) where Braun is located and where Dieter Rams lives. In Hustwit's imagination, the car would have the weathered, scarred, and forgotten patina of a long-forgotten race car.

While scouring the Internet one day, Hustwit came upon ROCS Motorsports on Instagram and started following the builder. He was intrigued by a silver 911 whose build had been inspired by the Fletcher Aviation 550 Spyder that had raced in La Carrera Panamericana Race in 1954. Based on that build, he believed ROCS and its owner Richard Conclaves was the right constructor for the car he had in his head.

Now, the hunt was on to find the proper donor car. Hustwit spent a year searching through ads, but each he saw had some sort of issue. Some were too rough and others too nice.

He came very close to purchasing a G Body Turbo Look 911, but the car was simply too good. It was a car he would be afraid to park anywhere. Then one day, Conclaves suggested Hustwit purchase his personal car for the project. It was a 1987 Carrera 3.2 that had been sorted and was ready to go. Further, it was now a 3.4-liter (207.5 cu in) car with several desirable suspension modifications. The perfect donor car had been right under their noses!

The following year was filled with creative decisions and sourcing the right parts. The hard work and long hours paid off. As you circle the car, it presents an endless array of subtle modifications inside and out. The rear deck lid looks as though it had been designed by Dieter Rams, its Glöckler grill badge making a complementary statement. The wheels look black, but when you look more closely, you realize they're dark green.

ABOVE LEFT: Subtle changes everywhere complete the package. 550 Spyder-style mirrors and a flush gas cap with no filler door are unexpected.

ABOVE RIGHT: Stripping the cover off the dashboard revealed the crude metal casting lying beneath. When you look closely at the oversized oil temperature gauge (OEL-Temp), you'll realize that the letters are created with adhesive, giving the effect of a sticker worn off or removed.

From the hood-mounted CIBIÉ driving lamps to the subtle use of a side mirror from a 550 Spyder, everywhere you look there are fascinating details. The car's paint is a masterpiece of patina with ghosted layers revealing what would have been former race sponsorships or stickers.

You can't stop looking at the car because every time you go back you find more—the same with the interior. Inside, the exposed WEVO short shifter looks much better without the boot as does the steering wheel hub. "It just looks better stripped down," says Hustwit. The car originally had power windows, but these were converted to hand cranks to be more build-period correct.

A thorough examination of Hustwit's car leads you to appreciate the artistry created by Conclaves. It all makes perfect sense when you learn that he is an artist and a painter when he is not working on cars.

Perhaps the best thing about this build is the fact that underneath it all the car is a well sorted kick-ass 911 3.4! Hustwit says, "It really drives like an old race car," something he insisted upon throughout the build process.

When you consider the car in context to Hustwit's obsession for details—whether old guitars or brilliant design—the car makes perfect sense. Don't be surprised if you see another 911 from Hustwit soon. He has another five ideas spinning in his head right now.

CLOCKWISE FROM UPPER LEFT: (1) 911 R taillamps have almost become "standard issue" when backdating a Porsche Outlaw. They always look correct. (2) Finding the perfect hood mounts for the CIBIÉ rally lamps became a bit of an obsession for Gary. The hunt was well worth it. (3) The original Braun logo was created in 1934 by Will Münch. The logo is presently in its 4th iteration. The one used here was created in 1995 and is still being used today. (4) The placement of an Otto Glöckler dealer badge on the rear deck lid is another nod to a time long past. The Frankfurt, Germany, dealership was also known for its modified Porsche specials.

ABOVE: Lurking on the back streets or in a dimly lit alley, this Porsche Outlaw looks and feels most at home. With the glass exhibiting the only reflections, there is nothing shiny about it.

RIGHT: The mastermind behind the Kronberg Porsche, Gary Hustwit, is an independent filmmaker specializing in such topics as industrial design, typography, and graphic design.

JEFF ZWART—A DIFFERENT KIND OF OUTLAW

Jeff Zwart's outlaw tendencies were passed on from his father, Bob. Zwart was brought up in the back of his father's 1964 Porsche 901, chassis number 35, and he learned to drive in that car. Zwart's father didn't leave his Porsche factory stock. It was just an old Porsche when he purchased it in 1968—why not upgrade things?

He installed 7-inch (18 cm) 911 R wheels, a 1968 911 L engine, aftermarket exhaust, and suspension parts from the *Porsche Sports Purpose Manual*. Basically, Zwart Sr. was adding things to his car based on the theme of *Sports Purpose*. In fact, Freeman Thomas and Jeff joke today that Zwart's father may have been an R Gruppe member before there was an R Gruppe. Today, based on the value of a 901, you would never think of altering an original. With everything period correct and original, that car would be worth over $1 million today.

Zwart is an R Gruppe member, and his 911 is what inspired the club's creation. Rather than modifying an early 911 to include parts from the *Porsche Sports Purpose Manual* (essentially a guideline on how to convert a 911 to a 911 R), he owned *the* holy grail: a 1967 911 R, chassis number 11 (which is why Zwart's racing number is always #11). The first R Gruppe newsletter featured his car on the cover.

Zwart's passion for Porsche and driving is synonymous with his name. He's raced everything from a 914/6 in the Panama to Alaska Rally to the TransSyberia Rally in a Cayenne S to a 935/19 at Pikes Peak.

Zwart has driven just about everything Porsche has ever made, so I was curious about his Porsche 906. The 906 was built as a race car—the first time Porsche had used fiberglass in a car's construction. Unlike the preceding 904, which was a road and race car, the 906 was intended primarily as a *race* car. However, there is a loophole. The 906 could be licensed and considered street-legal in all countries except Germany, where it was a designated race car only. The 906 has turn indicators, windshield wipers, and other items you need for the street.

"It's really kind of fascinating that in just three short years starting in 1966 Porsche transitioned from the 906 to the 907, 908, the 910, and then the 917 in 1969," Zwart observed. "So, in just three years they went from a [flat-six] 906 to a twelve-cylinder 917. Talk about ramping up your motor sports program! That was probably as aggressive as you can get."

Zwart has also raced the 2018 Porsche 935/78. There were only 77 of these cars built, and Jeff has driven one on the racetrack and in a Time Attack race at Pikes Peak in 2020. "It's a funny race car to drive because underneath all that beautiful carbon fiber and incredible design that pays homage to the 91 standard road car," Zwart said. 'It's not often you get into a fully caged, stripped-out race car where you have to put it in park, put your foot on the brake, and start it with a key. It's pretty unusual."

The 935/78 is essentially a Porsche GT 2 RS underneath all its fancy body cladding. Zwart describes the original 935 as "kind of an outlaw. They sloped the nose of the car, then Kramer put in two back windows to meet the rules. You have one window at one level, which was the actual factory window, then another window above that to give it better aerodynamics. There was a lot of that kind of outlaw racing-style thought process brought to the cars that allowed Porsche to win well into the '80s with the 911 based car."

NEXT PAGE: Jeff Zwart paying tribute to his late friend Ken Block at Luftgekühlt 9. After thrashing the throttle of the BBI Autosport-prepared 1400 horsepower *Hoonipigasus*, Zwart's heartfelt words touched the hearts of everyone in the room.

X
BYE
LF 4304

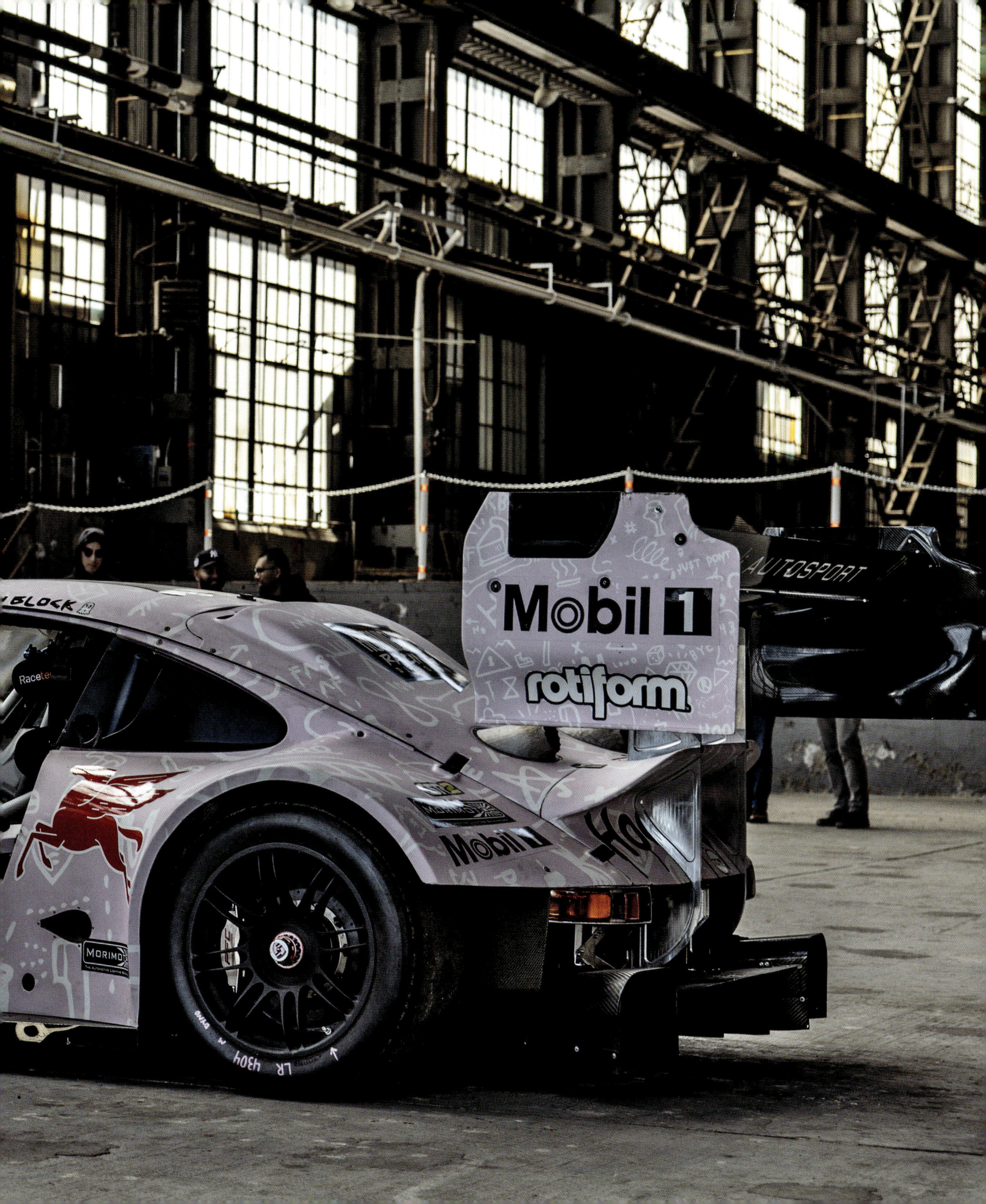
Mobil 1
rotiform
AUTOSPORT
Mobil 1

One of Zwart's Pike's Peak race cars being prepped at E-Motion Engineering in Costa Mesa, California. Careful attention to ride height, performance, and safety equipment is critical when running flat out on gravel.

Zwart has also had the opportunity to drive all kinds of Porsche Outlaws built by both private and professional builders. He has the utmost respect for cars developed by Alois Ruf, Singer Vehicle Design, and Rod Emory. But when it comes to his own cars, he explains, "I want to drive things that offer the experience of what it was like exactly in that day. I don't need 200 horsepower [147 kW] in a 3.6. There's nothing wrong with [that approach]. It's just that I like it when a car dictates your pace, and I find that I really enjoy how my world slows down. I look at that as being something very special." Zwart describes his mindset as being "pure to the brand. I'm actually extremely careful that everything is exactly as it was in that period."

There are very few people who take Zwart's approach to their Porsches. When you see him driving a 906 up California's interstate 405 to go to Cars and Coffee or brandishing studded snow tires and a canoe on the roof of 1953 Gmund coupe to drive through fresh snow to a lake in Colorado's twisties, you start to realize that he is a different kind of outlaw. He *drives* cars that are period correct. Zwart's Porsches aren't resting in a hermetically sealed vault. He takes them out and experiences them in the way they were intended to be experienced. Zwart's car doesn't have to be the fastest on the road. He can do that any time he wants on the racetrack. Instead, he enjoys the period correct experience of every car he owns.

ABOVE: Jeff behind the wheel of his all original 1965 911 Coupe as he stretches it's legs on California's SR 1.

LEFT: From above we can see exactly why Jeff's 1953 Gmünd Coupe is so unique. The combination of the split windshield and wrap around front quarter windows offer an interior greenhouse like no other.

BOTTOM: Jeff is no stranger to Dawn Patrol at Amelia Island. Here he is in 2015 with his 1971 Porsche 914/6 GT Werks Monte Carlo Rally car.

ABOVE: Joshy's car is no stranger to long hard drives throughout the west. Occasional stops just to take it all in are a must.

OPPOSITE, TOP: Oil stains, mud, and bugs are frequent hitchhikers and welcome on Joshy's travels.

OPPOSITE, BOTTOM: One of Joshy's T-shirt designs for Luftgekhült. "Know the Ledge" is a take off of how Joshy lives and encourages you to drive your Porsche the same way.

JOSHY ROBOTS

I first met Josh Coburn (a.k.a. Joshy) on a chilly spring morning in Emeryville, California, at the monthly EASY Cars and Coffee gathering.

His 1969 911 stood out in the crowd based on the fact his white car has a red passenger door. That door has become its signature as well as his. It's a statement: "Hey, the door works. Who cares what color it is?" The door perfectly matches Coburn, a nonconformist with no intention of ever painting that door to match the rest of the car.

Coburn purchased the Porsche as his only car after becoming frustrated living in San Francisco and renting Zipcars to go snowboarding. "I've always been a driver."

He found a ridiculous rust bucket in Walnut Creek, California. Apparently, its owner had found the car on Craigslist in South Carolina. The car arrived, and the owner discovered the real story after pulling up the carpet and finding there were no floors in the car. He could see through to the ground below. The owner was fine with it and knew he could fix it. His wife didn't agree, so the car went back on Craigslist where Coburn found it.

Coburn's brother has been rebuilding 1960s Ford Mustangs for decades, so Joshy sent him photographs of the floors. His brother assured him it could be fixed. That's all Coburn needed to pull the trigger because he knew beyond the rust the car was a good runner.

"Every time I fix something on the car now, I just fix that [one] thing," Coburn explained. "I don't get involved with cosmetics etc. It just needs to run." Of course, one day his foot did go through the floor pan. He shipped the car to his brother who slapped new floors in it. Done.

Apparently, Coburn is known for blowing up engines on rallies and having a new engine in the car 30 days later for the next rally. "I like finding, buying, 'Frankenstein-ing,' and getting back on the road," explained Coburn. "I see myself more as a driver, and I can't drive without a car!"

Coburn is known in the community for his brand Joshy Robots, which he describes as a "manufactured vessel for ideas that spill out when you're a creative person, living a particular life, having the privilege to adventure old Porsches as I do."

Joshy Robots creates everything from cool T-shirts to glazed-donut grill badges to replica 917 gear shift knobs made out of broken skate decks. The knobs sell out as quickly as he can make them.

Coburn is outspoken and creative. I sat down with him to find out how he feels about the term *outlaw.*

"I'm a writer and somebody who feels like words have meaning. Even when you ignore their meaning they have meaning. And I think the community, the Porsche world, has abused the word *outlaw.* It's a very important word. At least it was once. It carried a lot of weight which is why people chose to use it. They used it with intention."

We talked about other terms like *restomod* and how it might even be more fitting for what most people do these days. But the term *restomod* is, in Coburn's view "pretty milk toast—not badass enough. Using the word *outlaw* is basically just someone trying to sexy up the term *restomod.*"

Coburn and I talked over a long cup of coffee, discussing semantics and even debating whether Thomas Kincaid's work was art or not. We both agreed it was not.

We discussed the criminal implications of the word *outlaw,* and we eventually stumbled upon what Joshy referred to as an aha moment.

"Outlaws don't call themselves outlaws," Coburn noted. "We label someone an outlaw and the irony [is] people taking on the label themselves."

Just like a hot rod, there are all kinds of definitions—from *rat rods* to *street rods* to *customs* and *low riders.* It seems we're conflating and confusing terms because at one moment we're talking about the car and the next we're talking about the individual who built or owns it.

I wondered, if some of the cars being built today and featuring tons of cosmetic work but very little in the way of performance modifications should be described as Porsche Outlaws or considered fashion?

Coburn's eyes rolled followed by a long pause. He then said, "Fashion is super complicated. Part of fashion is lovely, and part of fashion is art, but also part of fashion is robbing aesthetic from aesthetic, and the question is: Do you put so much energy behind that, that we totally starve the authentic expression?"

This discussion and cross referencing trying to define the meaning of the word *outlaw* can go on indefinitely. Ones expression of this word through their build is as unique and varied as the outlaws in this chapter.

CIBIE
CIBIE
CIBIE

No "Mall Crawler" here! Jim Goodlett is known for thrashing his way everywhere he goes.

7

SAFARI CARS

ABOVE: E-Motion Engineering owner Joey Seely is known for his calm demeanor and confident approach. It consistently assures his customers they have come to the right place (that blur in the background is Jeff Zwart's Pike's Peak race car).

OPPOSITE, TOP: One of the latest projects from E-Motion is this off-road ready 911 Turbo S. Why would you do this? Because you can.

OPPOSITE, BOTTOM: Take a second look at the driving lamps integrated into the front bumper and the brush bar neatly designed to follow the original body lines.

JOEY SEELEY AND E-MOTION ENGINEERING

From the moment you enter E-Motion Engineering, you know builds are done to a level completely different from other builders. Stepping into the office feels like you've just entered a Day Spa. Everything is orderly and in its place. Even the air you breathe has been curated, just like a five-star hotel, with a particular scent filtered through its ventilation system to create an emotion and to signify the essence of the owner's message. That may sound crazy, but it's true.

From the front of the shop to the back of the shop, this feeling is consistent. None of this is by accident. It's all on purpose, or should I say purposeful, because E-Motion Engineering believes that any client should experience something completely different from any other shop they've been in.

This is Porsche legend Jeff Zwart's favored shop. In fact, it occupies what was Zwart's former office and garage space before he decamped to Colorado. E-Motion is not too far from the former location of Porsche Motorsports North America (PMNA) and ANDIAL in Costa Mesa, California. The workspace glistens with shiny clean floors and white walls. The shop is uncluttered, and there are no posters, photographs, product banners, or stickers of any kind on the walls. The space is much more like standing in a gallery or museum environment. Lighting is subtle and indirect, creating a pristine atmosphere. What customer wouldn't want to have their car in that environment? It's like choosing the finest school for your gifted child.

Now, add in decades of hands-on racing experience, as well as engineering and race-support experience. Owner Joey Seely has truly done it all.

Everything changed for Seely when he helped Patrick Long and Jörg Bergmeister in 2005 and 2006 when they were racing a 2006 Petersen Motorsports White Lightning Porsche 911 GT3 RSR. This effort was only the second time in history that a car run by a privateer had beat Porsche Motorsports in a FIA World Endurance Championship while using the company's own car. Since that point in time, Porsche has made it a point to know exactly what Seely is up to at any given moment.

Seely is also famous as the chief support team behind Jeff Zwart's multiple successes in the Pikes Peak International Hill Climb (PPIHC), as well as supporting numerous other drivers including Rhys Millen and Tanner Faust.

I first met Seely in November 2014 to photograph a car called *Project Nasty*. That car was a revolutionary, hot rod Porsche Outlaw. He was working with BBI Autosport at the time.

After leaving BBI in 2015, Seely's first project was designing and building Luftauto Number 001, which was auctioned by RM Sotheby's at Luftgekühlt 3 to raise money ($275,000 in this case) for the Autumn Leaves Project charity. Number 001 was a 1985 911 Carrera Coupe converted into a rally-style car. The car was so admired that there have been several knockoffs of its design. Periodically, people approach Seely, point to a Safari 911 and ask, "Hey, is that one of your cars?" Smiling, he explains, "No, but it could've been. I can't say that we started a [Safari] trend or that we own it, but it's certainly flattering when you see people copy your work."

911 turbo S

E-MOTION
911 turbo S
FALKEN

ABOVE: Here's a great example of the wide range of Porsche cars found at E-Motion on any given day. There's everything from classics to overlanders to future rally cars.

OPPOSITE: A closer look at the driving light setup without the bumper in place reveals this is not an afterthought by any means. It's all well designed and executed to look like it was done at the factory.

Some of the Safari cars Seely has been commissioned to build could go out and break competition records. But he also knows that some of his builds just cruise to Cars and Coffee. Owners simply enjoy the dual-function potential of their Seely-built cars.

With GT cars, he knows his customers like to drive a car to the track, race all day, and then drive it home and go out for dinner that night. "That's what we pride ourselves on," Seely said, "the duality of our vehicles."

Currently, GT cars are the majority of Seely's business, and he has so much experience building them he can essentially turn them around in a week's time. The Overland/Rally cars, however, require more time and planning. But regardless of what the team is working on, Seely says, "It's still a hot rod mentality, but instead of American muscle it's Porsches."

Working with a new customer on a build is an intense interview process, Seely explains. "We need to understand their goals, what [is it] they want [to] achieve? It's important to know if someone uses the vehicle as a daily driver or wants to go to the track occasionally. From there I call out the modifications that will be done."

Seely's two least favorite words are *corner balance* and *alignment*.

Most people think these are all-encompassing terms. "It's not," he says. "That's like saying, 'I had a steak.' You can have a $15.00 steak, and you can have a $100.00 steak. Our work is all about tailoring to your needs. Corner balance and alignment are most important because we are maximizing everything about the car and the components you have to [suit] your driving ability and what you intend to use it for."

E-Motion has moved beyond the 911 and is now offering Cayenne Overland builds as well. Five packages and suspension stages are available based upon customer needs and expectations. Add-ons include just about anything you can imagine from winches to light bars, custom skid plates, and more.

The Cayenne builds are a growing side of the business that Seely thinks will continue to expand with endless possibilities. E-Motion has yet to build a Macan version, but that will happen as soon as his fiancé lets him have the keys to hers.

While visiting the shop, I had the opportunity to photograph a new 911 Turbo S they'd just completed with an off-road package. The car included suspension mods, skid plates, and tubular corner protection plus full under-chassis protection done in Kevlar. Driving lamps integrated in the front bumper are reminiscent of Seely's signature through-the-bumper exhaust featured on his *Project Nasty* car.

Equally trick was a GT3 receiving a full rally conversion, including a spare wheel mounted inside the car with access via an operational rear window reminiscent of the 997 Targa's rear window opening.

I asked Seely for his thoughts on Porsche's new 911 Dakar cars. "They've done something smart," he said, "by making a limited number of only 2,500 vehicles. It may actually help me." Not to mention the fact that the 911 Dakars are going for considerably more than sticker. Seely's conversions seem like a bargain and offers more functionality.

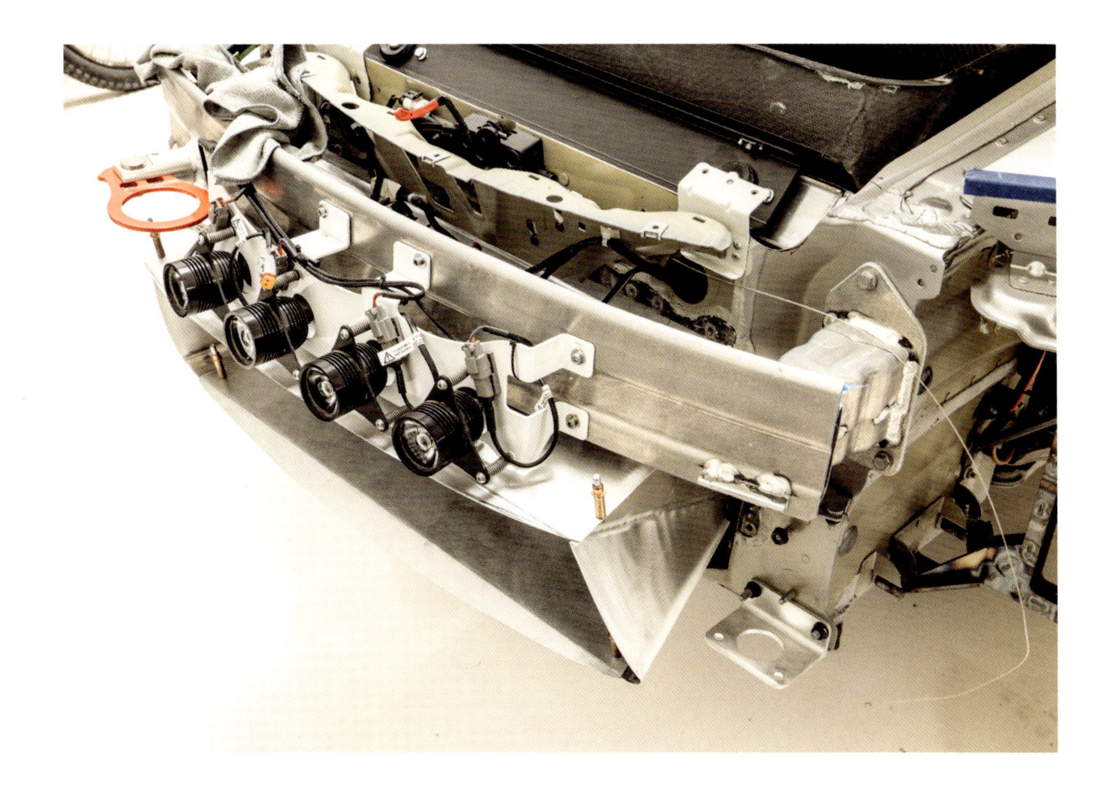

RIGHT: Where else would you see a new Porsche GT3 Touring car getting a full-blown rally setup? Note the full-size spare mounted under a custom hinged rear window design similar to that of the 997 Targa.

BELOW: This is not your typical mall crawler. E-Motion's customers are hard-core and known for demanding the best equipment available. From suspension upgrades to snorkels and undercarriage plates, E-Motion does it all.

LEFT: Integrating a full-size spare into a roll cage under glass is no easy project. Here's a closer look at how it's done.

NEXT PAGE: Synonymous with Seely's name is the car that put him on the map. *Project Nasty* was the test mule for everything going forward. It looked great then and still looks great now.

As mentioned earlier, Pikes Peak is a huge effort for E-Motion these days. The planning stage starts as early as August of the year before the event with design and then full-on construction and prep from January right up 'til race day in June.

Chassis tuning for Pikes Peak is complicated because it's basically three courses in one. You have the flowing bottom section, followed by tight twists in the middle including 13 first-gear hairpins, followed by an extremely bumpy summit.

Addressing those tuning challenges, Seely brings vast knowledge of different types of racing to bear. His experience all comes together, offering anyone working with him a strong competitive edge.

His mechanical skills are enhanced by having had the opportunity to work with some of the best drivers out there, benefitting from their behind-the-wheel feedback. Other members of the E-Motion team bring extensive racing knowledge of their own to the party.

Seely says the sum of his experience allows him to "look at the matrix of the car—in motion—not standing still on a lift or on the ground. It allows you to visualize the movement of the wheel and the suspension arms and where the weight is placed inside the axles, outside the axles, up high and down low. You start to see all the dynamic aspects of the car. This helps you imagine the car on anything from a loose surface to operating at the limits of adhesion."

Keep a close eye on E-Motion Engineering (just like Porsche does). Its experience and potential are exceptional. I wouldn't be surprised if history repeats itself and Seely and E-Motion become the next ANDIAL.

AUTOSPORT
RECARO

775

A lineup of Resolute Motorsports's builds—from street car to rally car to Overland Cayenne.

SCOTT BIRDSALL

Scott Birdsall is known for shaking things up. He's a proud "Hoonigan," with an extensive racing history. But he is probably best known for his antics in a $225 1949 Ford F-1 Craigslist purchase turned into an unconventional race truck. Known as *Old Smokey F1*, it was powered by a Freedom Racing 6.7-liter (408.9 cu in) Cummins diesel compound turbo motor delivering 1400 horsepower (1,030 kW) and 2,100 pound-feet (2,847 Newton meters) of torque. If that's not wild enough, it also holds the diesel record at the PPIHC. (Unfortunately, the truck was totaled in 2023.)

When Birdsall isn't racing, he runs his shop, Chuckles Garage, in Santa Rosa, California. The shop is known for its ability to fabricate anything you can imagine for your project. On any given day at Chuckles Garage, you can see everything from Mustangs to Japanese tuner cars to old-school hot rods to Birdsall's Overland Porsches from his Resolute Motorsports project.

Resolute Motorsports is Birdsall's brand for specialized Porsche Overland 911 and Cayenne Overland builds. He's adamant that his builds are meant to be used.

"They are designed from a motorsport standpoint, and everything is built to take whatever off-road abuse you can throw at it," Birdsall explained. "I'm not some poser slapping body panels and shit on a Porsche."

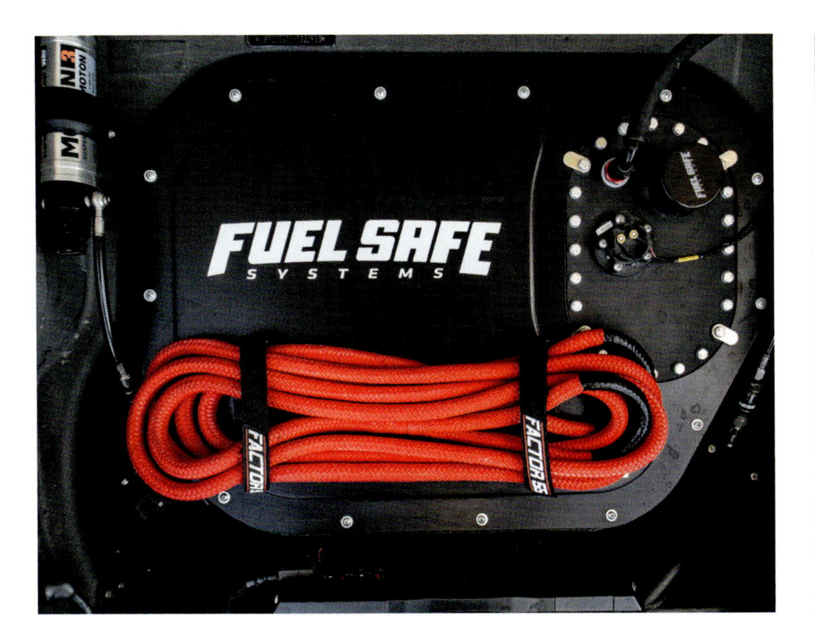

ABOVE LEFT: Going the extra mile for safety is what it's all about in a car that can go anywhere. The standard fuel tank has been replaced with a Fuel Safe Systems tank for added safety.

ABOVE RIGHT: This is not your typical 911 instrument cluster. Note this MoTeC C127 cluster includes the Resolute Motorsports logo and that the tach is still the largest instrument dead center—just where you would expect it in a Porsche.

LEFT: Here's the business end of a Resolute 911 Overland complete with a Rasant Intake System installed.

Resolute vehicles come in four stages of build based on the customer's needs. The Cayenne that Birdsall drives every day is a 2014 Cayenne diesel with a stage-four build including all the bells and whistles. It began life with all the factory options except air suspension. Resolute worked with a company called Eurowise to develop a tubular suspension system that would retain the Porsche factory magnetic ride system, the Porsche Active Suspension Management (PASM).

MOTON
PORSCHE

Stage-four builds are further equipped with 33-inch (84 cm) tires on 18-inch (46 cm) ALPHA Wheels, roof rack, rear-tire rack, tuned exhaust, light bars, and other accessories.

The 911 Overland Resolute builds are focused on function and not mere appearance. Birdsall explained that "I can't help but come at this from [my] motorsports engineering background."

Resolute Motorsports worked directly with Moton Suspension to develop dampers specific to their 911 and Cayenne builds.

Birdsall's favorite donor car for a 911 Overland Resolute build is the SC. "There's just something about a G50," he said. "It's just a good car."

Build turnaround times depend upon your preferred level of planned vehicular chaos. Stage one cars can range from 60 to 90 days. A full restoration plus Overland conversion can take a solid year.

Birdsall loves driving a Resolute 911 on the road after punishing it and getting it filthy off-road. "I'll drive it around like that for weeks. Some people are stoked to see it, and others are just dumbfounded." Purists may believe he has ruined the car. To them his answer is: "I made it, and I made it more useful for what I want to use it for. It's my car and I'll do whatever the hell I want to it!"

I asked Birdsall what he thought was next for the rally craze. He seems to think the serious rally people will always be there, but those who are just into the "fashion" aspect will fizzle out. In fact, he thinks the craze is beginning to trail off. He noted that ever since Porsche announced the 911 Dakar, the demand for his stage one and two cars has dropped.

Of course, when you consider 911 Dakar pricing, one could be driving a stage-three Resolute car for similar money. But people seem to prefer the convenience of a turnkey, factory-built car. "Fashion and convenience win again," notes Birdsall. Still, he'd like to see 911 Dakar try to keep up with a Resolute car in the desert. "That would be a fun day!"

LEFT: Scott Birdsall is dead serious when it comes to racing and performance. He doesn't just build things. He races them flat out on a regular basis.

OPPOSITE, TOP: A Resolute Overland Safari car rests patiently before its next round of thrashing the North Coast of California.

OPPOSITE, BOTTOM: Rear angle shows how the undercarriage plate protects the exhaust with its punch through design.

ABOVE: Jim Goodlett has lived his life on the edge since he was a kid, and he's still a kid at heart. He proves it every day living life *slideways.*

OPPOSITE, TOP LEFT: Form *does* follow function. Jim placed the oil cooler just beyond the fan inside the ventilated plexiglass window. Heat is exhausted through the plexiglass window's holes.

OPPOSITE, TOP RIGHT: There's no stopping Goodlett's Cayenne from getting all the air it needs with the help of an E-Motion Engineering carbon fiber snorkel. Puddles and creeks be damned!

OPPOSITE, BOTTOM: Where the business gets done. Notice the lack insulation and sound deadening materials . . . hence the headphones and headset for communication purposes dangling from the roll bar.

NEXT PAGE: This is Goodlett's current stable of off-road Porsches. Each is ready to go at a moment's notice.

JIM GOODLETT "SLIDEWAYS"

Jim Goodlett's irreverent attitude and approach to life is summed up in one word: *slideways.* It's no accident or marketing ploy. Goodlett has been living his life slideways for decades. It's an approach encompassing everything from skateboarding and surfing to road bikes, mountain bikes, and motorcycles (both road racing and dirt), and on to Porsche 930 Turbos, 911 SCs, Cayennes, and now a Porsche 356. I guess you could say any day going slideways is better than any day you're not.

Goodlett can talk about going slideways for hours, citing endless examples from his vast experience on many different platforms. But to sum it up, he explains that "Slideways is about finding the edge where the wheel catches, the suspension catches, and all of those things come together. You don't even have to be the fastest and the greatest—there are plenty of people faster and more accomplished than I will ever be. It's about finding where the edges are. It's a dance of chaos, but when chaos happens you throttle through."

Goodlett and I discussed the "Scandinavian Flick" at great length. It's an old trick where you intentionally get a car out of shape and then snap it into a reverse lock to slide through and return control. He refers to this as "controlled chaos." You're essentially throwing the car out of balance, first to get it to go in the other direction. "Off-road you want that car upset," Goodlett explains, "and when the [steering] wheel goes to the opposite lock you're really controlling it by the rear wheels."

This all started for Goodlett at the age of four at the dinner table. His father who "didn't have two nickels to rub together" would talk about driving his friend's Porsche 356 and 550 Spyder at Watkins Glen. Lengthy conversations about this company called *Porsche* and how innovative it was made a further impression on young Goodlett.

In time, he got hold of some Hot Wheels cars and track, but unlike his buddies, his racecourses were not built in a circle or a straight line. Goodlett had a different take on things. His tracks would run into a wall, a bookcase, or maybe out the door and onto the steps just to see what would happen. This was the birth of slideways.

Then one day, Goodlett and his father were watching the International Race of Champions (IROC) on television. There were twelve Porsches on the same track, all matched vehicles from the factory, thus creating an even playing field for twelve top drivers. "They were these bizarre machines with these massive tails on them and they were racing flat out," Goodlett recalled. Suddenly, the two lead cars connected in the chicane and both went off track. Instead of breaking into pieces, the cars were undulating across the terrain like nothing Goodlett had ever seen. He found himself wondering, "How on earth are these cars doing this and nothing else you would see from NASCAR or Formula 1 would ever be able to withstand this?" His dad explained, "It's a Porsche. It can do anything. That rear-engine and rear-wheel drive gives them an unheralded ability to have traction and balance as the gas tank is up front. That huge wing in the back is all about downforce."

It was at that moment that Goodlett knew he would "one day" own a Porsche.

A few years later in 1978 when Goodlett was 16, a buddy returned from England with a car magazine featuring an article about the hardest race on the planet:

SLIDEways
E-MOTION
OZ

MARTINI
522
BILSTEIN
BILSTEIN

EXIT
VREDESTEIN
837

ABOVE: Perhaps the most iconic of Goodlett's Rally cars, the 522 doing what it does best—throttling through an off-road turn at speed. Driving the 522 like this requires the right setup and all the parts . . . not to mention the frame of mind and sheer guts to thrash your Porsche mercilessly.

RIGHT: Jim has even applied his slideways attitude to his Cayenne Diesel, the 958, which continues the theme of occasional wash downs only done by splashing through the mud or sprinting in the rain.

the East African Safari Rally, which consisted of six days of racing and ranging from 5,000 to 7,000 kilometers (3,107 to 4,350 miles). After reading the article and learning how Martini Racing had convinced Porsche to build cars for its team, Goodlett was more convinced than ever that he needed a Porsche and soon.

But "soon" was followed by 28 years of tire kicking and smudging his fingerprints on the windows of numerous showroom cars including a memorable 1984 gray market 930 Turbo on a Houston used car lot in 1984. This *Widowmaker* had been imported by a local doctor who determined in short order that the car was too much for him. He instead bought a new 911 SC from the local Porsche dealer. Because the '84 930 was a gray market car, the dealership could neither sell it nor take it in trade. As a courtesy, the dealer allowed the doctor to put it on their used-vehicle lot. They passed the doctor's business card to whomever might be interested, and eventually the 930 sold.

That Turbo had made such an impression on Goodlett that in 2012, at age 50, he found himself on the hunt for a 1984 930 Turbo. Not just any 930—he wanted one matching the one he'd spotted in Houston so many years ago. As luck would have it, a car surfaced matching his description. In fact, it turned out to be the exact car Jim had seen in Houston back in 1984. Kismet could not be denied, and Jim bought that car as his first Porsche—but certainly not his last Porsche.

In 2014, Goodlett decided to build his own rally car based on a 1978 911 SC. The build was based on the spec of the Martini cars run in the East African Safari Rally. The car is fully caged with some modifications such as struts from Elephant Racing. While the car was being built, Goodlett spent two to three hours a night for 45 days "practicing" his off-road driving on Xbox!

When his tribute car burst on the scene it was a hit. Goodlett even drove it right onto the lawn of the PCA Werks Reunion event at Amelia Island in 2016 fully dressed in as much mud and as many bugs as he could find along the way from Savannah, Georgia. Rebellious!

Continuing with the off-road theme, Goodlett decided to build a Cayenne Turbo Diesel based on the Porsche 955 Cayenne S TransSyberia rally vehicle. It's a worthy tribute, considering Porsche's first through sixth place sweep in the 2009 rally.

The Cayenne build has proven itself off-road where Goodlett reports that it drives like a heavier 911 but can attack water features a 911 would never be able to handle.

As it turns out, the 356 has become Goodlett's favorite Porsche. "I will tell you, if I had to get rid of all of them but one, the 356 would be the one I would keep and here's why: Its sense of going *slideways*. It's a different kind of slideways because it's not mid-engine and it's not rear-engine. The gas tank is [located] more forward making it uniquely balanced."

Goodlett says "It's no wonder you see so many Porsche 356 cars through the years at the Rally Monte Carlo. It's the perfect car." Bear in mind this Rally changes from tarmac to gravel and is run in everything from ice to snow.

What's next for Goodlett and his slideways theme? Well, I wouldn't be surprised to see a 912 somewhere down the road. It's the car Goodlett and I both agree offers a perfect package, balancing design with the simplicity of adequate performance.

RIGHT: The 958 may be larger than the 522, but its setup defies physics when it comes to its agility off-road.

BELOW: Resting peacefully on the border of Georgia and South Carolina, this dynamic duo doesn't appear winded despite the thrashing they just received before arriving here. As the sun falls, we have plenty of light to guide the way.

(continued on page 164)

SLIDEWAYS STABLE

Jim Goodlett seems to be making up for all those years he waited to get that first Porsche. Safari, Rally, and Outlaws in his garage today include the following:

- 1974 Porsche 911 RSR build featuring a heavily reworked 3.0-liter (183.0 cu in) engine with performance cams, high-compression pistons, PMO carbs, straight headers, and a fiberglass body. The complete car with Goodlett at the wheel weighs just 1,868 pounds (847 kg).

- 1978 Porsche Safari Rally 911 SC racer with CIBIÉ Super OSCAR triple auxiliary lights, Heuer stopwatches, fully caged, modified suspension, modified air intake, MOMO seats and safety harness, and Stilo communication headsets. The car was built to the same spec as the 1978 Martini racing Porsche 911 SC cars that campaigned the treacherous East African Safari Rally race that same year.

- 1982 Porsche Safari rally 911 SC built to nearly the same specification as the 1978 rig but with RECARO seats instead of MOMO.

- 1984 Porsche 934 with MOMO Supercup Racing seats, SCHROTH harness, Rennline floors and foot kick panels, period correct CIBIÉ rally lights, Stilo race headset and communication gear, Terratrip Rally computer, Heuer rally stopwatches, and an engine with a type 935K 27 Turbo.

- 1987 Porsche 944 Turbo (951)—bone stock.

- 2015 Turbodiesel Cayenne with E-Motion Engineering lift kit and carbon fiber snorkel, BILSTEIN shocks, OZ Racing wheels, BF Goodrich KO2 tires, four Eurowise aluminum skid plates, and a 10,500 lumens light rack to blind the commoners when advancing towards them at 105 miles per hour (169 kph) on a mountain fire road.

- 1963 Porsche 356 B T6 tarmac rally car with a Steve Hoffman-built 356 C motor. The motor's original 60 horsepower (44 kW) has been doubled. Engine modifications include a performance cam, higher compression pistons, Solex carbs, and custom velocity stacks for improved air flow.

ABOVE: Expect the unexpected, they said—anything goes at the Baja 1000. That's exactly what Chris Harrell and Mike P built, and the results are killer.

OPPOSITE, TOP: Mounting all the rally lights on the bumper eliminated the issues involved with creating a designated light bar or finding a hood-mount system.

OPPOSITE, BOTTOM LEFT: The Porsche crest provides the perfect center point from which all other options are measured and mounted.

OPPOSITE, BOTTOM RIGHT: From BRAID wheels with knobby tires to cool graphics, the purpose-built Baja 924 S has it all.

(continued from page 162)

BAJA 924 S

This whole project started with an advertisement on Instagram. One day in April of 2018, the phone rang. The voice on the other end said, somewhat cryptically, "Hey, I signed us up for the Rally XL. It's a rally in Baja early next year. I hope you're in. By the way, we'll need to build a car."

Meet Mike Prstojevich (a.k.a. Mike P) and Chris Harrell. Based in Portland, Oregon, the best friends and coworkers have been involved with various product design and manufacturing projects for decades. Following that call, the next big decision was: "What car are we going to drive?" The rally was ten months away and they were starting at zero—not to mention that neither of them had ever competed in an off-road rally or developed a car to do so.

Based on other life distractions, the car search didn't start until mid-summer 2018 and the build process didn't commence until September. That left two rookies four months to build a rally car for a 5,000 kilometer (3,107 mile) event south of the border at the end of January in 2019.

After multiple texts, cups of coffee, impromptu conversations, DMs on Instagram, and considering vehicles from a used Mercedes-Benz G-Wagon to a Mitsubishi Pajero, they were still empty handed and seeking a set of keys for the right vehicle.

But Prstojevich and Harrell are both Porsche fanatics. Prstojevich owns a backdated Rothsport Racing air-cooled hot rod 911 and Harrell has owned everything from a 914 to a 928 to a short wheelbase, long-hood 911 to a few different Cayennes and a 991—currently over 20 Porsches in all. A Porsche seemed to make the most sense for sticking with the theme.

Now, they had several things to consider: budget, modifications, timing, reliability, and terrain. Most everyone suggested they build a G-body 911, but the cost of those cars was skyrocketing, and anything within budget was unrealistic when it came to reliability.

So, they zeroed in on a Porsche 944. Transaxle cars are much more affordable and accessible. Then, Prstojevich stumbled on some images of a 924 that was raced in Monte Carlo. That looked pretty cool, but they were still fixated on a 944.

They tossed around the idea of a transaxle car with several mechanics. One by one each of them concluded, "You guys should do a 911." Even Jeff Cameron from Rothsport pled, "Please, please don't bring me anything but a 911."

But Prstojevich kept recalling cars he had seen in Stuttgart at the Porsche Museum. One of them was a black 924 Carrera GTS with those flared fenders, wide tires, and rally lights up front.

ABOVE: Resting peacefully in the bucolic setting of Nicasio, CA, after thrashing hard all day, the Baja 924S has proven its ability to get the job done and get you where you need to be.

OPPOSITE, TOP: A post-Baja romp in NorCal before heading back to Oregon offers a great test for wet conditions.

OPPOSITE, BOTTOM: Ripping through the wet pastures of Sonoma provides additional insight to the capabilities of this build. The car proved unstoppable.

ABOVE: Here they are showing off the 924's capabilities upon arrival in Mexico for the Baja 1000. CHRIS HARRELL

OPPOSITE, TOP: Snagging a moment of rest to take in the sunset before staring the night run to the next rest point in Baja. CHRIS HARRELL

OPPOSITE, BOTTOM: Once a thriving bodega, this shell of a building now offers little more than a place to find shade from the intense Baja sun. CHRIS HARRELL

As fate would have it, Harrell found a Guards Red 924 S on Craigslist in the summer of 2018. The owner was asking $3,000 for the car, but a test drive revealed engine-smoking issues as well as a few cosmetic challenges. Further, the previous owner had purchased the car just to use the VIN number to register himself in a Porsche-owners only golf tournament. They were able to get the car for $2,200.

Surprisingly, for their $2,200, Prstojevich and Harrell ended up with a car that was obviously some past PCA member's pride and joy. The 924 had some deferred maintenance issues, but the car was really clean and just needed some TLC.

The more they thought about it, the more Prstojevich and Harrell got into the idea of the 924 S as rally car. In retrospect, even though they thought they had to find a 944, they ended up with something far more rare and that was only built in 1987 and 1988.

THE BUILD

In building the car, they replaced the nose and the hood with 931 Turbo parts, which included inlets for improved airflow. The front and rear bumpers are OEM that Prstojevich drilled out.

The car had to look cool, so they started designing a wrap that would be eye-catching. While they were considering wrap options, they took time to address a few body imperfections like road rash and what appeared to be an impact with a guard rail at one point. Those sections of the car were repaired and repainted. Someone else would have simply placed the wrap over any imperfections and called it a day. Not these guys. Harrell stated, "We wanted to do right by the car [and] we wanted the underlying car to be nice."

Prstojevich found himself fixated on a light pod for the hood. He had seen something on the hood of a Hugo Boss car in Europe, but he couldn't find it. He tried to develop something with 3-D printing. After several attempts, Prstojevich ended up calling Leh Keen and purchasing one of his light kits. Unfortunately, it did not fit the 924's hood. Finally, they bypassed the entire idea of hood-mounted lights and simply bumper mounted them.

They installed a 944 short-shift kit and a limited-slip differential (LSD) to help with any traction issues they might encounter. As it turned out, that limited slip was a key major mechanical mod that proved beyond useful.

As for the interior, a Porsche Classic radio was installed, the seats were opened and restuffed with new foam, and a GPS system was added.

The car then traveled to Yakima, where the company created a one-off fit using one of its roof-rack systems. This would allow Prstojevich and Harrell to carry RotopaX fuel containers on the roof as well as traction recovery boards in the inevitable event they got stuck.

At the time they were building their racer, there were very few bolt-on items designed for a 924 off-road setup. Custom skid plates were created for the front of the car, and the back and front ends were raised two inches (5 cm) using custom spacers.

Underhood, the only issue was a leaky manifold gasket. A new timing belt and clutch were installed. No additional modifications were made to the engine.

BRAID wheels mounted with BFGoodrich KO tires would handle traction duties, and mud flaps were installed to keep debris at bay. Out front, a winch was mounted to handle extractions.

OREGON
SP 28928

ASFOLT
BAJA XL
18

Schedules didn't allow for anything more than a quick shakedown run around the back woods outside Portland, Oregon, before being loaded on a transporter. Luckily, everything seemed to check out because the next time they would see the car would be in Baja the night before the Rally.

BAJA XL

Keep in mind that this event, the BAJA XL, is not sanctioned by the National Off Road Racing Association (NORRA).

From the event website: "The BAJA XL is open to 'anyone by anything.' If it's street-legal you can drive it. The rally is open to cars, motorcycles, trucks and whatever else you can think of."

Not surprisingly, Prstojevich and Harrell's 924 did not experience any extensive scrutineering upon arrival. In fact, they found themselves well equipped and ready for the most competitive class. "It was fairly serious and just fun," according to Harrell. Checkpoints were transponder based. There was something for everyone depending on how serious you wanted to make it.

After a while, they decided to become their own kind of outlaws and scrapped the transponder and just enjoyed the drive and the amazing topography.

Prstojevich thought the first day would be an easy stage to get acquainted with the course and ease into things. They were to end the day at Mike's Sky Ranch and have a nice dinner. Instead, the day proved challenging, and they didn't arrive at the destination point until 10:00 p.m. after 13 hours on the road.

The next morning, after being handed the map that included some rock crawling in the first stage, they knew they would be more comfortable doing their own thing again.

It was all great until they got stuck in silt for most of the day, 20 miles (32 km) from anything. Their winch proved useless in this case as there was nothing solid to attach it to—no other vehicles, no big trees, only low brush and cacti. But that's what the adventure is all about, right? They eventually extracted themselves and got back on course.

A couple nights later, as they pulled into Todos Santos north of Cabo San Lucas, they noticed a tear in a lower radiator hose. Luckily, their hotel was significantly nicer than any other inn they had stayed at along the way. They were "stuck" a second time, but this time the accommodations were more pleasant.

They spent the next day shuttling between a local shop and an AutoZone in Cabo San Lucas trying to find the right hose. With no luck, they took the car to a shop in El Pescadero, where they flushed out the system and reattached the leaky hose after mending it with Gorilla Tape that Harrel had tossed in the car at the last minute before leaving on the trip.

Needless to say, the trip South of the Border had been "hosed," and they headed home. Along the way, they stopped at Race Service in Los Angeles and then attended the first Morning Shift event at Porsche Experience Center in Carson, California, the following day. Their racer was a total hit!

After the dust had settled, literally and figuratively, and the road had been washed from the car and themselves, they both agreed they would definitely run the event again—and in the same car! Though there were some stages in which the car was not competitive, it proved to be an excellent balance of power, handling, and reliability.

OPPOSITE: This shows the typical terrain for the Baja 1000. What car doesn't look good surrounded by cactus and scorched desert scrub?

ASFOLT
BAJA XL
18

MADE IN THE
USA
HOOD RIVER, OR
CHAMPION
C1
C2
C3
SPACERS
RICAL
EMORY
MOTORSPORTS

Following in her father's footsteps, Jayde Emory knows exactly where every nut and bolt is in the shop.

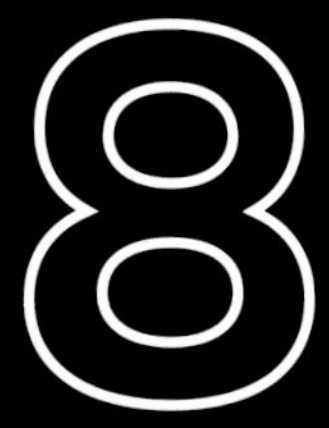

THE OUTLAW FUTURE

The Outlaw Porsche movement has been motoring on for more than a half century and no one's looking for it to slow any time soon—on the contrary, it's exploded over the past ten years.

There was a time when no one outside the fashion industry knew the name Magnus Walker, Porsche Outlaw influencer extraordinaire. Now, he's had an exhibit at The Petersen Automotive Museum in Los Angeles celebrating his Urban Outlaw brand and featuring 10 of his Porsche Outlaw cars, his books, swag, and graphics.

Rod Emory created his custom-build and restoration empire, Emory Motorsports, in 1996 and is known the world over. Emory's brand has gone from a small dream to an icon of excellence and creativity. The current wait for an Emory Outlaw extends years, and his client list reads like a who's who of Porsche royalty. When Hurley Haywood, one of the winningest 24 Hours of Daytona drivers, and rock 'n' roll legends like car guy John Oates order your cars, you're doing something right. Emory's extensive referral list says it all.

Luftgekühlt started as a local experiential Porsche car happening in Venice Beach in 2014—a gathering of like-minded Porsche enthusiasts of all ages and from all walks of life from L.A. hipsters to gray-haired enthusiasts. It's the brainchild of Porsche factory racer Patrick Long and creative director Howie Idelson and ably curated by Jeff Zwart, world-renowned photographer and cinematographer. It's beyond local now, with Porsche fans flying in from all over the country to experience it.

All of this activity suggests that while it takes a certain kind of individual to appreciate the outlaw ethos, the ranks of those who choose to express themselves through a hot-rodded Porsche are clearly growing.

Personalization is the root of hot-rodding, whether it's performance-related or cosmetic. This obsession is the common thread that connects this community.

There are many drivers who happily settle for an automobile rolled off the showroom floor in a color they can bear for the length of a lease. Maybe it's a lack of imagination, or maybe it's just a different set of priorities.

I do know that *they're* not reading this book.

The past several years have seen Porsche Outlaws cropping up in other countries and on atypical platforms. Enthusiasts are throwing caution to the wind. The more radical and outlandish the better, as long as it's done right. Builders dropping $400,000 into a 928 or constructing a Baja/Rally conversion of a 924 S, turning a former grocery getter into a serious off-road contender—why not?

The possibilities are endless, and all kinds of companies are jumping into the Safari game.

Rob Dickinson and his Singer Vehicle Design built the unexpected Singer ACS and flipped the industry on its ear. Clearly, Singer has more up its sleeve than just restomodded 964s and 911 Turbos.

Equally surprising, The Keen Project out of North Carolina is now up to its 30th Safari car build. What started as a wild idea has now turned into a business for Leh Keen and shows no sign of slowing.

As for Porsche AG, all this enthusiast energy helps expand its reach and encourages more and more people to support the brand. No other major manufacturer is embracing alternative projects quite the way Porsche does.

ABOVE: Luftgekühlt started as a local event in the parking lot of Deus Ex Machina in Santa Monica, CA. It has grown into a curated destination event for the Porsche crowd.

LEFT: Patrick Long's passion for the Porsche brand knows no bounds, and it shows no signs of slowing any time soon. As Luftgekühlt expands to include water-cooled cars with its Air|Water events, who knows what's next?

OPPOSITE, TOP: A rare moment, Magnus Walker rests easy on a break from his grueling schedule as the most prolific proponent of Porsche Outlaw culture.

OPPOSITE, BOTTOM: The future is bright for Rod Emory and Emory Motorsports. He is living his life, surrounded by his family and establishing a legacy of Porsche Outlaws.

As Porsche continues creating outlaw-like concepts that trickle down into the market, we have no need to be concerned about it stopping. PORSCHE AG

In 2023, Porsche announced its 911 Dakar, a tribute to the original 959 rally car that Walter Röhrl piloted in 1986. Röhrl performed the testing for the new 911 Dakar, allowing the story to come full circle. Porsche's past success with the 959 and more recently with the Cayenne TransSyberia are both examples of the company's marketing savvy and response to its customer base.

And, true to Porsche form, they let builders and the aftermarket test the market and then come in with what I like to call "a better mousetrap."

The fact that a customer can now go into any Porsche Dealer and order a Safari car or something like a GT2 or GT3 is a sure sign of encouragement for future Porsche Outlaws.

We will start to see more restomods accepted as part of the mainstream.

The days of making sure your car is as original as possible are slowly moving to our rearview mirror and instead will include motor swaps, brake upgrades, handling modifications, and the like.

A perfect example of this is a car created with the Porsche Classic Department and commissioned by the PCA. Called the *Classic Club Coupe*, this junkyard 996 was restomodded with one-off body panels, a ducktail, twill Pepita-patterned seat centers, and even a 996 GT3 Mezger engine. The car is completely custom from nose to tail.

Further proof that restomods are gaining traction in the Porsche world is PCA's new event called ÜnStock, celebrating modified Porsches only.

As we drift towards the internal combustion engine (ICE) horizon, inspiration will come from the ever-evolving merger of technology and imagination. Future demands will encourage more electric conversions of classic Porsches as enthusiasts get more comfortable with the idea.

Will the conversion of a long-hood, air-cooled 911 into an electric powered car with flared fenders and rear deck louvers work for the R Gruppe? Probably not, but if they are cool enough, they might at least avoid being shunned. After all, Freeman Thomas, a former Porsche designer and R Gruppe cofounder, is now producing the electric Meyers Manx, the original dune buggy retooled for the twenty-first century.

Porsche Outlaws are here to stay—all signs point to it. A younger audience seems drawn to the movement as well.

While attending media day for the Luftgekühlt 8 in 2023, I overheard a female voice behind me proclaim, "I never want to get out of this car." I was pleasantly surprised to see that this was a young woman who was being photographed in the event shirt. She was referring to Patrick Long's 1966 912, a repainted, rough around the edges, plain-Jane Porsche with some Benton Performance upgrades and black race wheels.

Her genuine enthusiasm speaks to the strength of the Porsche Outlaws' continued success and is a welcome sentiment to a die-hard, gray-haired car guy like me.

Kayla Delehant is a prime example of where this trend is headed. Her passion for the sport and brand is contagious.

Clay Carr is carrying the outlaw torch at full throttle. He's turning his childhood obsession into reality and is one of the youngest members of the R Gruppe.

CLAY CARR

Because we enjoy the Porsche Outlaw subculture, we can't help but wonder where the next group of supporters will come from. Time and again, mainstream media suggests that younger generations have diminished interest in cars and driving. How will that impact the car hobby and the Porsche Outlaw world specifically?

Relax. Meet Clay Carr. At 27, Carr is exactly what the hobby needs.

He's young, smart, and he drives the wheels off his Sepia Brown 1973½ 911 T. He also owns Scott's Independent in Anaheim, California, a Porsche service, restoration, and repair shop specializing in long-hood 911s.

The car world is part of Carr's DNA. His family has been in the car business for decades in California. His family roots are based in truck sales, supplying lettuce farmers with all kinds of Fords to help keep agriculture moving in central California.

Carr grew up in Orange County, and he has been attending Cars and Coffee events as far back as he can remember, as well as events like Donut Derelicts in Huntington Beach and shows in Costa Mesa, Irvine, Laguna Beach, and Crystal Cove.

Carr bought his first car, a Volkswagen Beetle, at 18. He had zero mechanical knowledge at the time, but he loved the car and the community that supported it.

Owning your own car means working, both on the car and at a job, to keep it on the road. Carr's first job was at a little shop restoring Fiat Jollys, just around the corner from his parents' house. It was a somewhat challenging gig for the 6 foot 6 inch (2 m) Carr, who can scarcely fit in a Fiat Jolly. He cleaned parts, working on a barter basis in exchange for cool parts and accessories for his Beetle.

After high school, Carr entered Pepperdine University, where he majored in business. When home for the summers, he worked at a shop called Carparc USA, in Costa Mesa, California, which specializes in short wheelbase Porsches. Carr describes the owner, Henk Baars, as a "standup guy in the industry and just the nicest, most honest guy you could ever want to work with."

Carr started at Carparc deploying the same skills he'd used in the Fiat Jolly shop, namely disassembling and cleaning parts and components. This work provided the opportunity to begin visualizing how things worked together mechanically. Carparc USA introduced Carr to the historical side of the industry because at that time, owner Baars was creating a database for the early 911 cars. Carr spent hours inputting VIN numbers, part and engine numbers, option codes, interior colors, and paint colors. He would also input information on who owned various cars and what historical significance a given car may have had. "I love history," Carr explained, "and I learned the importance of the provenance of certain vehicles based upon who owned them, who raced them, where they came from, and where they were now."

The experience solidified Carr's passion for early 911s. He had always been a car guy, "but it was there [at Carparc], in those moments, that I became a Porsche guy."

After another year at Pepperdine, Carr returned to Orange County and took a job at Scott's Independent in Anaheim, California. Clay and the shop's owner, Scott Henry, had known one another for years through the Cars and Coffee scene. As it turned out, it was the perfect opportunity for Carr as Henry had plenty of

time to mentor him because the shop was small and not focused on turning out volume work.

Over the next three months of Carr's summer break, both Henry and Dan Reese tried to cram 50 years of mechanical knowledge into Clay's brain. By that fall Carr said, "It was enough to know I really loved [the business]."

Both Henry and Reese are R Gruppe members and rather than leave Carr alone at the shop, they began dragging him to events and to the track.

After an extensive vetting process over the course of four years, Carr was asked to be a member (his R Gruppe number is 806).

After graduation from Pepperdine and after four years of working at Scott's Independent, Carr is now the shop's proud owner.

As a shop owner and enthusiast, Carr has a lot invested in the game. I asked him where he thought the Porsche hobby was going and what it might look like down the road.

He says he's getting more and more calls from people who have recently inherited a car or have just found the car they remember their father having when they were young.

If these walls could talk . . . Clay knows the history and stellar reputation of Scott's Independent is something he has to live up to every day. But that's exactly why he bought it.

ABOVE: Clay's Sports Purpose-prepped Sepia Brown 911 is not the typical car driven to the grocery store in SoCal.

OPPOSITE, TOP: A personal touch on Clay's car is the aluminum bumper guards. These guards create a 50 percent weight savings over the originals.

OPPOSITE, INSET: Clay just happened to have one of the R Gruppe grill badges with the *911* in what used to be the member number area. Lucky guy. The R Gruppe was asked by Porsche AG to remove this from the badge as it was a licensing issue.

Inheritance is one thing, but purchasing a vintage 911 is ever more challenging in that the entry fee has climbed dramatically. There was a time when you could buy a good used Porsche for $10,000. Now, a decent car is going to cost $100,000 or more. "I think the key to the future [of the hobby] is just giving people an opportunity like Scott and Dan did for me. Do everything you can to get someone into the seat, behind the wheel, and I think the money and the entry point becomes less of a hurdle. To experience the joy of a small-displacement engine, and what it can do. Because the current trend is bigger, better, faster, and yet a really happy 2-liter [122.0 cu in] engine is a wonderful thing."

Carr feels that there are many things happening to keep younger people involved in the Porsche brand as well as the outlaw scene. Social media, online resources, and YouTube all play major roles. He points out how fashion has relied on YouTube and Instagram for years. "At seventeen I can't go out and buy a 911, but I can experience other people's passion and plant a seed through merchandise and videography." There's nothing wrong with wearing a hoodie with a vintage Porsche graphic on it even if you don't have the car in your garage. Becoming interested or attracted to Porsche doesn't necessarily have to be about people buying cars. Creating exposure to the brand is important to the hobby as well. One day, that same person will get a big bonus check or save up for years to realize the dream. If someone sets a goal to one day own a Porsche Carr says, "Even if it's thirty years from now, we'll be here to help them."

(continued on page 184)

CARR'S CAR

Carr's Sepia Brown 1973½ 911 T was previously owned by a mechanical engineer who worked in the aerospace industry. It was serviced at Scott's Independent from the day the shop opened in 1974.

IN THE 1990s, the previous owner burned out the motor and took it upon himself to rebuild it. After blowing up the motor a second time, he realized that though he was a mechanical engineer, he certainly wasn't a Porsche mechanic.

Scott's Independent bought the car in the late 1990s, and Henry built it as an outlaw. He ran it with the R Gruppe for several years, but eventually it ended up parked on the side of the shop.

Soon after Carr came to work at Scott's, he had his eye on the T. He worked out a deal with shop owner Scott Henry to purchase it and build his own Sports Purpose car.

Carr built a 3.2-liter (195.3 cu in) long block engine with early 911 SC injection resulting in "massive torque."

Young Carr wasn't interested in comfort features and immediately ripped out the air conditioning and anything else that added weight or served any purpose other than performance. Exit the stereo and amp and so long huge sport seats, replaced with a fiberglass bucket out of a 911 S for him and a stock passenger seat for his wife. "She's really short, and I wanted her to be able to see over the dash-board," he explained, justifying the extravagance.

Multiple engine upgrades later, Carr now describes the T as a "pretty well sorted hot rod." It's traveled all over the United States and currently shows 496,000 miles (798,235 km) on the clock.

There's something about these green dials that pull the instrument cluster together.

Accommodating his wife with a more fitting seat shows what kind of guy Carr is. He's smart. You know the saying: "Happy wife, happy life."

Cameron Wayland is an old soul. He has knowledge and a temperament way beyond his years. His understanding and passion for Porsche is on another level.

(continued from page 181)

CAMERON WAYLAND

Not long after interviewing Clay Carr, he called to suggest I speak with Cameron Wayland, another young enthusiast. I looked up Wayland and discovered that his shop, Wayland Motorsports, in Sebastopol, California, was less than an hour from my home.

Upon arrival, I realized I had probably passed his shop a dozen times or more in the last few years never knowing what was inside. Set off Guerneville Road in an unassuming red and white aluminum building marked only by a small white sign, it's easy to blow right by.

That's a reflection of the beauty of Wayland Motorsports. There's nothing pretentious about the shop, and everything is matter of fact. I arrived early but was soon greeted by the sound of an approaching air-cooled 911 breathing through those distinctive Solex carbs.

As it turned out, Wayland was driving his own car that day. His 1966 911 is finished in Irish Green with a brown interior and real RECARO seats that were a factory add-on back in the day. The car offered a few other clues that it had once been set up for Sports Car Club of America (SCCA) racing, things like attachment points for a roll bar, which was no longer in place.

The car showed up one day at Wayland's shop via a referral from a friend who had rebuilt the car's Solex carbs. Wayland had had a conversation with the Washington state–based owner who just decided one day to trailer the car through the night and drop it off at Wayland's shop.

He and the owner hit it off and the two spent a great deal of time going through the car to create a list of what the owner wanted done and to ensure the client would be satisfied once the work was complete.

Unfortunately, the owner's available budget was significantly less than the quote. After several discussions, and even considering the option of selling the car on the owner's behalf, Cameron told the owner how much he loved the car and how he wanted to become its new owner. They came to an agreement and Cameron has been driving the car ever since.

The previous owner has since passed away, but Wayland still hears from the owner's wife periodically. She is pleased that Cameron is carrying on the legacy of *Greeny* as she calls the 911.

This story is typical for Wayland Motorsports. Authentic stories with soul are connected to every Porsche passing through its doors. Some stories are better than others of course, but nevertheless continue to be personal and real.

The story of the shop itself is another one of those tales.

After Wayland opened the shop in 2016, folks began showing up when they saw Porsches parked outside. Many had stories about how they had serviced their Porsches there in the 1970s and 1980s. Apparently, the shop was once called Tillman's German Car Specialists. "I had multiple cars I had never seen before come by," Wayland said. "I would open the [car's] door and it would have an old Tillman service sticker inside the [jamb]—from this same building and this same location. Cars had stickers from the building I am working in from before I was born."

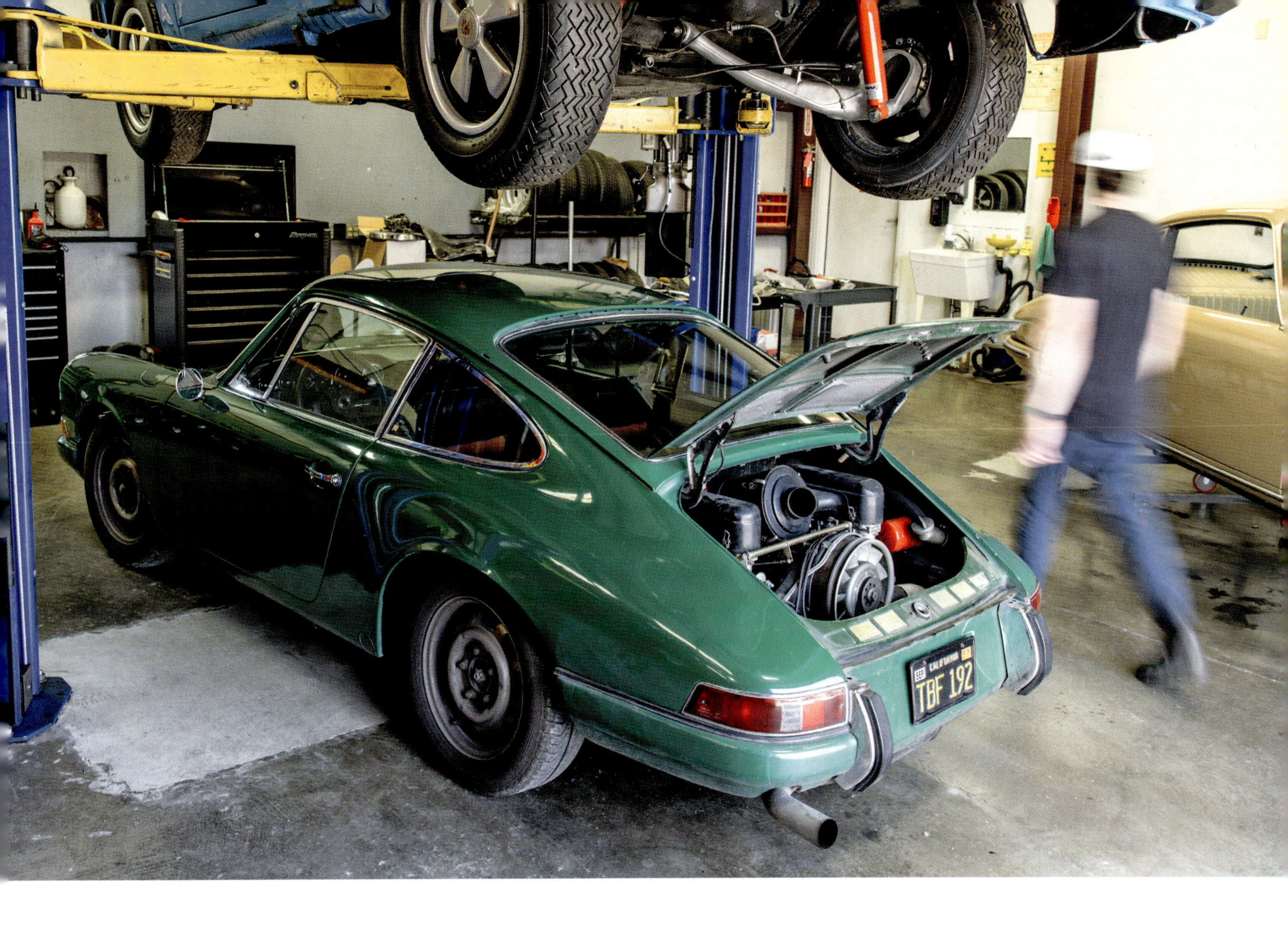

When Cameron isn't driving his 356, you'll find him behind the wheel of this Irish Green '66 911.

Wayland says his whole business started organically after taking his Porsche on spirited drives through Sonoma County backroads with friends. Prior to the 911, Wayland had a 1972 Datsun 240Z and had become part of a great enthusiast community. When he found himself with a little extra cash, he purchased a 1975 911. That purchase introduced him to a whole new sports car community. Eventually, he rebuilt the 911's engine. "You've got a fast car—what did you do to it?" friends would ask. One by one his enthusiast friends started asking if he could do a similar setup on their car.

Wayland's business approach is what keeps customers coming back. Clients approach him all the time with lengthy lists of mods they want done. It would be easy (and profitable) to simply do what they ask and take the money. But Wayland feels that the original *Porsche Sports Purpose Manual* makes a lot more sense than most of the things his clients have in mind.

Before he starts making modifications on a customer car, he instead suggests, "Let's go for a drive. You drive what you have now, and I'll drive something that is basically stock and underpowered and you see if you can hang with me. Then we'll talk about upgrades."

ABOVE LEFT: Steel wheels with no fancy hub caps are another sign the car is "all go and no show."

ABOVE RIGHT: Race belts and sports seats are a dead giveaway of the attitude of this Porsche Outlaw—not to mention the steering wheel.

RIGHT: The car is a treasure chest of original parts and patina. Don't let it fool you . . . you can tell by the stance it's dialed in.

More often than not, people simply disappear in his rear view mirror. Once the drive has stopped, they ask "What the hell is going on?" Wayland responds with "Let's get this thing dialed in and as close to stock—like the way it came for the factory—before we start talking about upgrades."

Wayland believes in getting every car dialed back to factory specs and its owner driving quickly before he'll start making the sort of modifications he has done to race cars and other builds. There's nothing like the right set of tires, shocks, alignment, and corner balance to show what the car was is capable of.

One of Wayland's biggest concerns is just how much longer enthusiasts will be able to enjoy driving their gas-powered performance cars in California. "There seems to be a time limit to this experience," he worries, "which is why I'm all gung ho about trying to get people out and driving things instead of just buying them. Before we know it, just like horses, you'll be able to own one [gas-powered car], but you'll only be able to drive it on your own property, not downtown."

Looking forward, it seems to me the Porsche Outlaw craze is in good hands for a long time to come.

ABOUT THE AUTHOR

MICHAEL ALAN ROSS spent years photographing automobiles and people (both commercially and editorially) before one brand snapped into focus more sharply than any other: Porsche.

That clarity came courtesy of a 1979 Porsche 911SC he purchased from Joan Jett's drummer. At that moment, Michael's passion for cars and photography collided, launching a journey that continues to this day.

In *Porsche Outlaws*, Michael melds his knowledge of European and classic cars with his California hot rod roots to yield a deeper understanding of the Porsche outlaw form and the culture of those who build, own and, most importantly, *drive* them.

In addition to his latest Porsche work, Michael has shot for clients including Porsche Design, Porsche Cars NA, Polestar, Axalta, Ford, and Opel AG. His photography and editorial work has appeared in outlets like *BBC Top Gear*, *Guitar Aficionado*, *Road & Track*, *Automobile*, *Hot Rod Magazine*, *Hot Rod Deluxe*, *The Rodder's Journal*, *Hop Up*, *EVO*, *Excellence*, *000 Magazine*, *Porsche Panorama Magazine*, *Christophorus*, and book publisher Motorbooks. He and his wife, Danielle, reside in the San Francisco Bay area.

https://michaelalanross.com/

INDEX

Special thanks to my dear friends Joey and Sue Shimoda for their help and support.

COLORFUL
24
52-640
19 COLORADO 53